Trade and Climate Change: Focus on Carbon Leakage, Border Carbon Adjustments and WTO Consistency

Other titles in Foundations and Trends® in Microeconomics

Natural Gas Pipeline Regulation in the United States: Past, Present, and Future
Matthew E. Oliver
ISBN: 978-1-68083-452-9

The U.S. Market for Uranium: 70 Years of History
Charles F. Mason
ISBN: 978-1-68083-382-9

Game Theory and Water Resources: Critical Review of its Contributions, Progress and Remaining Challenges
Ariel Dinar and Margaret Hogarth
ISBN: 978-1-68083-016-3

Trade and Climate Change: Focus on Carbon Leakage, Border Carbon Adjustments and WTO Consistency

ZhongXiang Zhang
Founding Dean and Distinguished University Professor
Ma Yinchu School of Economics
Tianjin University, 92 Weijin Road
Tianjin 300072, China
ZhangZX@tju.edu.cn.

Founding Director, China Academy of Energy
Environmental and Industrial Economics
Tianjin University
Tianjin 300072, China

now
the essence of knowledge
Boston — Delft

Foundations and Trends® in Microeconomics

Published, sold and distributed by:
now Publishers Inc.
PO Box 1024
Hanover, MA 02339
United States
Tel. +1-781-985-4510
www.nowpublishers.com
sales@nowpublishers.com

Outside North America:
now Publishers Inc.
PO Box 179
2600 AD Delft
The Netherlands
Tel. +31-6-51115274

The preferred citation for this publication is

Z. Zhang. *Trade and Climate Change: Focus on Carbon Leakage, Border Carbon Adjustments and WTO Consistency.* Foundations and Trends® in Microeconomics, vol. 12, no. 1, pp. 1–108, 2018.

ISBN: 978-1-68083-482-6

Foundations and Trends® in Microeconomics

Volume 12, Issue 1, 2018

Editorial Scope

Topics

Foundations and Trends® in Microeconomics publishes survey and tutorial articles in the following topics:

- Environmental economics
- Health economics
- Industrial organization
- Labor economics
- Law and economics
- Public economics

Information for Librarians

Foundations and Trends® in Microeconomics, 2018, Volume 12, 4 issues. ISSN paper version 1547-9846. ISSN online version 1547-9854. Also available as a combined paper and online subscription.

Contents

Trade and Climate Change: Focus on Carbon Leakage, Border Carbon Adjustments and WTO Consistency

ZhongXiang Zhang[1,2]

[1] *Founding Dean and Distinguished University Professor, Ma Yinchu School of Economics, Tianjin University, 92 Weijin Road, Tianjin 300072, China; ZhangZX@tju.edu.cn.*

[2] *Founding Director, China Academy of Energy, Environmental and Industrial Economics, Tianjin University, Tianjin 300072, China.*

ABSTRACT

This monograph provides a comprehensive review of the literature on competitiveness and leakage concerns associated with differentiated climate abatement commitments among countries. The literature reviewed is not exhausted, but it is sufficient to provide a balanced view of both academics and policy circles. Section 2 discusses main channels of carbon leakage. Section 3 discusses how to identify the sectors at a risk of carbon leakage. Section 4 examines *ex ante* estimates of potential carbon leakage rates, and explains why they differ from *ex post* results of environmental tax reforms and greenhouse gas emissions trading schemes that have been implemented in the European Union. Section 5 discusses broad policy options to address competitiveness and leakage concerns, and compares which anti-leakage policy, border adjustments or output-based allocation, is more effective to

ZhongXiang Zhang (2018), "Trade and Climate Change: Focus on Carbon Leakage, Border Carbon Adjustments and WTO Consistency", Foundations and Trends® in Microeconomics: Vol. 12, No. 1, pp 1–108. DOI: 10.1561/0700000060.

limiting carbon leakages or mitigating production loss in the sectors affected. Given that border carbon adjustment measures were incorporated in the U.S. proposed congressional climate bills to level the carbon playing field and could have potential conflicts with World Trade Organization (WTO) provisions and practical difficulties associated with their implementation, Section 6 discuses in great detail the WTO consistency, the effectiveness and methodological challenges of border carbon adjustment measures. The monograph ends with some concluding remarks.

1

Introduction

There is increasingly scientific evidence confirming man-made climate change and its resulting negative effects. The Fifth Assessment Report of the Intergovernmental Panel on Climate Change, the most comprehensive assessment of the science relating to climate change, reported with 95% certainty that the major cause of global warming was increasing concentrations of greenhouse gas (GHGs) produced by human activity (IPCC, 2014). Continued GHG emissions will cause further warming and have the potential to seriously damage the natural environment and affect the global economy, making it the most pressing long-term global threat to future prosperity and security. However, with greenhouse gas emissions embodied in virtually all products produced and traded in every conceivable economic sector, effectively addressing climate change will require a fundamental transformation of our economy and the ways that energy is produced and used. This will certainly have a bearing on world trade as it will affect the cost of production of traded products and therefore their competitive positions in the world market. This climate-trade nexus has become the focus of an academic debate (e.g., Bhagwati and Mavroidis, 2007; Brack *et al.*, 1999; Charnovitz, 2003; Copeland and Taylor, 2005; Ismer and Neuhoff, 2007; Swedish National Board of Trade, 2004; Zhang, 1998b and 2004; Zhang and Assunção,

2004; Zhang and Baranzini, 2004), and gains increasing attention as governments are taking great efforts to implement the Kyoto Protocol and forge a post-2020 climate change regime to level the carbon playing field (e.g., Parker and Grimmett, 2009; The World Bank, 2007; UNCTAD, 2010; WTO and UNEP, 2009).

Under the United Nations Framework Convention on Climate Change (UNFCCC) principle of "common but differentiated responsibilities," developing countries are allowed to move at different speeds relative to their developed counterparts. As a result, the Kyoto Protocol in 1997 to the UNFCCC drew a clear line between developed and developing countries. Developed countries had specific obligations to control their GHGs, but developing countries did not. This difference in climate abatement commitments could lead production of carbon-intensive products to move away from carbon constrained countries to non- or less carbon constrained countries. This could in turn lead to losses of employment and economic output, in carbon-intensive sectors of these more carbon regulated countries. The fears of competitiveness losses undermine the support for abatement policy in developed countries.

Though extensively used in the public debate, the notion of competitiveness remains ambiguous. Some scholars (e.g., Krugman, 1994) view this notion meaningless at the national level. At the firm level, competitiveness refers to the ability of a firm to maintain or even increase international or domestic market shares and profitability. A firm' competitiveness is influenced both by micro factors, such as cost structure, product quality, trademark, service and logistical networks, and by macro factors, such as exchange rates, trade rules and political regime stability (Baron and ECON-Energy, 1997). Environmental taxes and regulations affect a firm's competitiveness by changing its relative production costs. For example, if a firm makes intensive use of energy, *ceteris paribus*, then imposing an environmental tax will increase its production cost relative to those less energy-intensive firms in the short term. Thus, it would experience a decline in competitiveness, whereas less energy-intensive firms would obtain a relative cost advantage in the short term. The changes in relative competitive positions would lead to "winners" as well as "losers" from the imposition of an environmental tax. "Losers" might even threaten to relocate their business activities to those countries that have relatively lax environmental regulations, if such a tax were put in place. This raises the question: do environmental taxes and regulations hurt firms' competitiveness so badly that they are forced to move to pollution havens?

There is growing and mainly empirical literature on trade implications of environmental regulations (e.g., Copeland and Taylor, 1994, 2003 and 2005; Ederington and Minier, 2003; Ederington *et al.*, 2005; Jaffe *et al.*, 1995; Levinson and Taylor, 2008). Grossman and Krueger (1993), for example, have examined whether pollution abatement costs influenced the patterns of the U.S. bilateral trade and investment with Mexico and found that "the available evidence does not support the hypothesis that cross-country differences in environmental standards are an important determinant of the global patterns of international trade." Jaffe *et al.* (1995) review and analyze over 100 studies on the potential effects of environmental regulations on the competitiveness of American industry, and conclude that "studies attempting to measure the effect of environmental regulation on net exports, overall trade flows, and plant-location decisions have produced estimates that are either small, statistically insignificant or not robust to tests of model specification.". Ederington *et al.* (2005) find that the costs of environmental regulations affect trade flows depending on the extent to which manufacturing in the more regulated U.S. market (vis-à-vis developing countries' markets) is "footloose." For example, the competitiveness effects are smaller if the costs of transportation are high, or there are local agglomeration economies, or the significant plant costs of an industry. Moreover, other factors may impact the likelihood of relocating economic activity to economies with lower regulatory costs. For example, the availability of transportation means to consumers may impact the decision to relocate (Ederington *et al.*, 2005).

However, the aforementioned findings may not necessarily go to be the case of carbon prices in future (Zhang and Assunção, 2004; Zhang and Baranzini, 2004). Environmental regulations and taxes applied to date have been relatively modest, and they fall short of the levels needed to achieve the objective of keeping a global temperature rise well below 2 degrees Celsius above pre-industrial levels agreed under the Paris Agreement (UNFCCC, 2015). They could thus have significant implications for competitiveness. Given that many governments across the globe regulate energy prices and some domestic greenhouse gas cap-and-trade schemes are designed to suppress price changes in electricity markets, the extent to which whether a domestic climate change policy translates into higher energy prices will determine the impact of that policy on the industrial competitiveness in a given economy.

Clearly, the discussion in this monograph fits within a growing trade and environment literature, and falls into this broad context of environment-trade nexus. In the context of unilateral or uneven climate policies, the fears of competitiveness deterioration could be particularly problematic for developed countries of distinct "regional" character, like Australia, Canada, and the U.S., partly because their provincial (or state) governments under the federal system are vested with significant political authority, and partly because energy-intensive industries are not spread evenly throughout these countries. Therefore, deterioration in the international competitiveness of energy-intensive sectors, while potentially economically disruptive in any country, could impose regionally uneven impacts on these countries (Rose and Zhang, 2004; Rose *et al.*, 2006; Garnaut, 2008; Rivers, 2010).

Since greenhouse gases are the uniformly mixed pollutants, namely, one ton of greenhouse gas emitted anywhere on earth has the same effect as one ton emitted elsewhere, simply shifting production of carbon-intensive products from the carbon-constrained countries to non- or less constrained ones can reduce the environmental effectiveness of the regulating country's efforts. This phenomenon is referred to as carbon leakage.

Much of discussion on leakage issues in a Kyoto-type world is very much along these lines. Does it remain relevant after the Paris Agreement? Under the Paris Agreement, all Parties have committed to strengthen the global response to the threat of climate change by keeping a global temperature rise this century well below 2 degrees Celsius above pre-industrial levels and to pursue efforts to limit the temperature increase even further to 1.5 degrees Celsius (UNFCCC, 2015). Almost all countries have outlined their climate action plans known as "nationally determined contributions" (NDCs), to be implemented from 2020 and expected to be scaled up over time. Therefore, this world looks different from the Kyoto-type world, where developed countries had specific obligations to meet emissions reduction targets but the rest of the world had no obligations at all. The leakage concern is reduced somewhat after the Paris Agreement (Kortum and Weisbach, 2016).

However, the carbon leakage problem does not go away. A number of assessment studies suggest that GHG reduction targets outlined by more than 190 parties in terms of NDCs are very disparate in their ambition levels (e.g., Averchenkova and Matikainen, 2016; Höhne *et al.*, 2017). Averchenkova and Matikainen (2016) suggest that the six G20 countries lack overall framework legislation or regulation on climate change, and also are either behind on meeting their 2020 targets or have not set any. Höhne *et al.* (2017) show that the climate ambition level of China, the EU, and the U.S. varies, depending on the perspective taken. Examining six key policy instruments (carbon taxes, emissions trading, feed-in tariffs, renewable energy quotas, fossil fuel power plant bans, and vehicle emissions standards), Compston and Bailey (2016) find that climate policy strength also varies across the six biggest emitters by far – China, the U.S., the European Union (EU), India, Russia, and Japan. Consequently, these differentiated ambition level and fragmented climate policy will well lead to large asymmetries in shadow carbon pricing across countries. Such asymmetries will undermine the effectiveness of more ambitious national climate policy initiatives through carbon leakage as domestic emission-intensive and trade-exposed industries will relocate to countries or regions without or with only quite lenient carbon pricing (Böhringer *et al.*, 2017b). Moreover, obligations in the Paris Agreement are legally non-binding, and countries may not comply with their obligations because there are no sanctions for noncompliance. As a result, leakage will still be a concern after Paris (Kortum and Weisbach, 2016).

This monograph aims to provide a review of the literature on competitiveness and leakage concerns associated with differentiated climate abatement commitments among countries. The literature reviewed is not exhausted, but it is sufficient to provide a balanced view of both academics and policy circles. Section 2 discusses main channels of carbon leakage. Section 3 discusses how to identify the sectors at a risk of carbon leakage. Section 4 examines *ex ante* estimates of potential carbon leakage rates, and explains why *ex ante* estimates differ from *ex post* results of environmental tax reforms and greenhouse gas emissions trading schemes that have been implemented in the EU. Section 5 discusses broad policy options to address competitiveness and leakage concerns, and compares which anti-leakage policy, border adjustments or output-based allocation, is more effective to limiting carbon leakages or mitigating production loss in the sectors affected. Given that border carbon adjustment measures were incorporated in the U.S. proposed congressional climate bills to level the carbon playing field and could have potential conflicts with the World Trade Organization (WTO) provisions and practical difficulties associated with their implementation, Section 6 discuses in great detail the WTO consistency, the effectiveness and methodological challenges of border carbon adjustment measures. The monograph ends with some concluding remarks.

2

Channel of carbon leakage

Carbon leakage could be considered to be a kind of international externality (Markusen, 1975; Hoel, 1996). It is defined as the ratio of an increase in CO_2 emissions outside the countries taking domestic climate policies to a reduction in emission within these abating countries relative to their reference levels (Felder and Rutherford, 1993; IPCC, 2001 and 2007). Carbon leakage is mainly driven through two channels (IPCC, 2007; Reinaud, 2008b; Dröge *et al.*, 2009; OECD, 2003).[1] One

[1]There are another two possible channels, both of which have not been addressed much empirically. One is the international cleaner goods price channel. Carbon constrained countries typically increase the demand and thus international prices of goods used to reduce emissions. Consider the situation where the EU imports ethanol from Brazil to reduce its carbon emissions. This would lead Brazilian cars to rely more on gasoline, thus leading them to emit more (Quirion, 2010). Another channel concerns the leakage via interactions between policy instruments at different levels in the political system. For instance, in Europe the presence of EU ETS means that one Member State that introduces an additional policy instrument would lead to carbon leakage of 100% at the margin because total emissions are given by the sum of the national caps. Goulder and Stavins (2011) develop on this in more generic terms based on the co-existence of federal and state climate efforts in the US. They found that when the federal policy sets limits on aggregate emissions quantities, or allows manufacturers or facilities to average performance across states, the emission reductions accomplished by a subset of US states may reduce pressure on the constraints posed by the federal policy, thereby freeing facilities or manufacturers

is the competitiveness channel. Countries that commit to emissions control will bear higher carbon costs than the competing counties without similar commitments. This will make their products most costly than products from the latter countries. So in the short term, this will cut their exports and lead to more products imported from countries without similar commitments. As a study for Noway suggests, this could be particularly a problem for a small country (Bruvoll and Fæhn, 2006). Over the longer term, this will lead to investment and production shift to the latter countries. This competitiveness-driven shifts of demand and production lead to shifts in emissions from carbon constrained countries to countries without similar carbon constraints.

Another channel of carbon leakage is via the international fossil fuel channel (Felder and Rutherford, 1993). Carbon-constrained countries adopt stringent climate policies to cut their consumption of fossil fuels. This will push down the international prices of fossil fuels. These reduced prices will in turn induce an increase in fossil fuels in countries with less stringent climate policies, thus leading them to emit more. In virtually all applied economic models incorporating both channels, the larger part of carbon leakage occurs through this fossil fuel channel, rather than via the competitiveness channel (Babiker, 2005; Gerlagh and Kuik, 2007; Böhringer *et al.*, 2010; Burniaux *et al.*, 2008; Fischer and Fox, 2009). Fischer and Fox (2009) show that for most energy-intensive sectors the leakage attributable to shifts in production is estimated to be 18-38% of the total carbon leakage when the U.S. unilaterally implemented a carbon tax of US$50 per ton of carbon emissions in several carbon-intensive sectors, implying that the remaining 62-82% is via the international fossil fuel channel. This suggests that any anti-leakage policies can be effective in reducing global emissions only if they can mitigate leakage from the fossil fuel channel. If most of carbon leakage is via this channel, border adjustments or output-based allocation, which unlike the carbon price do little to change relative fuel prices, can only reduce carbon leakage to a certain extent. Böhringer *et al.* (2010) show that, in the case of the EU unilateral policy to cut its carbon emissions by 20%, these anti-leakage policies cannot reduce the leakage by more

to increase emissions in other states. This leads to serious emissions leakage and a loss of cost-effectiveness at the national level.

than 33%, when compared with the full auction scenario. This is mainly because most of the leakage is due to lower global fossil fuel prices and increased demands for fossil fuels in other regions.

No matter which channel carbon leakage occurs, leakage may be overstated or even the size of carbon leakage might be misstated if abatement resource effect is not considered. The abatement resource effect refers to that the taxed sector draws resources away from the other sector or country, which reduces their output and emissions. Baylis *et al.* (2014) show that if consumers can shift their purchases easily, leakage may be overstated in models that do not allow for substitution in production. If consumer flexibility is low compared to producers' ability to abate pollution by use of other resources, overall leakage may be negative. Zhang and Zhang (2017) further examine abatement resource effect on carbon leakage both theoretically and empirically, decomposing the abatement resource effect into the consumption substitution effect and the multiplier effect and deriving closed-form solutions for these leakage effects. When imposing a carbon price on the electricity generation sector in China, the decomposition analysis suggests that of the four effects (namely, the scale effect, the consumption substitution effect, the multiplier effect, and the production substitution effect) identified, such leakage is mainly through the production substitution effect and the multiplier effect (Zhang and Zhang, 2017).

The fears of competitiveness losses and of carbon leakage are the main arguments against stringent climate policies, and in favor of sectoral exemption from a carbon pricing or free allocation of allowances. Competitiveness and leakage concerns are also the motivations of border carbon adjustments proposals (e.g., Zhang and Baranzini, 2004; Stiglitz, 2006; Subcommittee on Energy and Air Quality of the U.S. House of Representatives, 2008).[2] In his opinion pages at *The New York Times*, the Nobel laureate Paul Krugman points out that carbon tariffs are "a matter of leveling the playing field, not protectionism." (Krugman, 2009) ... they make sense, done right. I agree on both the economics and the legal aspects." (Krugman, 2009b).

[2]By examining each of the design issues–which goods, which emissions, and which countries border carbon adjustments should apply, Kortum and Weisbach (2016) conclude that border carbon adjustments will impose high administrative costs and need strong welfare justifications.

bn-ê

e.g. = pour ex.

3

The identification of sectors at a significant risk of carbon leakage

A number of studies have quantified and compared the effects of carbon pricing on different industrial sectors to identify the sectors that are at a risk of carbon leakage. In so doing, two set of indicators are commonly used. One reflects carbon cost increase. Carbon cost includes both direct and indirect costs. The latter stem from a carbon cost mark-up for the production of electricity used by the sector. So the carbon cost indicator, or a value at stake, is defined as a ratio of the sum of direct and indirect carbon costs to the gross value added or turnover of a given industrial sector. Another indicator reflects trade intensity. It is the sum of imports and exports divided by domestic market or turnover plus imports. Hourcade *et al.* (2007) show that for the United Kingdom 23 sectors are expected to exceed either 2% indirect carbon cost increase or 4% combined direct and indirect carbon cost increase relative to the gross value added under a carbon price of € 20/t CO_2 and induced electricity price increases of €10/MWh, with cement, and iron and steel facing a cost increase of above 25%. But all the sectors with a value at stake of more than 2% (indirect carbon cost) or 4% (direct and indirect carbon costs) only account for 1.1% of UK GDP. With Germany having a higher than the EU average share of heavy industry, the value at stake reaches 2% of German GDP for the same carbon price (Graichen *et al.*, 2008). For the EU as a whole, energy-intensive, including power generation, cement, refining, iron and steel, paper and pulp, petrochemicals, glass, and aluminium plus refining, accounted for less than 5% of the EU GDP and an even smaller share of jobs in 2005 (Ellerman *et al.*, 2010).

Much the same is true in the U.S. For most goods, energy is a small portion of the overall cost, which means that the added carbon cost would be small relative to the total price. U.S. EPA (2009) shows that for 75% of manufactured output, energy expenditures are less than 2% of the total. Only 1/10 of the value of manufacturing has energy expenditures above 5%. At a price of US$15 per ton of CO_2, output would fall by 2% or less in 80% of cases (Aldy and Pizer, 2009). A few industries - metals, paper, chemicals, cement and the like are both global and profligate enough to be at risk, but accounted for just over 3% of the U.S. GDP in 2005 and less than 2% of its jobs (Houser *et al.*, 2008). Even the most vulnerable industries would shrink by 5% (Aldy and Pizer, 2009), and would not suffer the Armageddon that lobbying groups are predicting (*The Economist*, 2008b). These results are very much in line with the finding from an early study by the Annex I Expert Group on the UNFCCC (Baron and ECON-Energy, 1997). That study undertook a static analysis of the cost increases from a price of US$100 per ton of carbon on four energy-intensive industries (iron and steel, non-ferrous metals, paper and pulp, and chemical products) in the OECD countries. These sectors represent 3 to 7% of GDP and 1 to 4% of labour force. While the average cost increase measured as percentage of total production value differs among countries and across sectors, it is generally low (below 2%) except for Australia and Canada. All these analyses suggest that other factors affecting price levels may well dwarf the price effects of a carbon price, at least at the rates that are generally considered in these studies.

Taken together, identifying the sectors at a risk of carbon leakage requires a combination of a minimum level of energy or greenhouse gas intensity (or cost) and a minimum level of trade exposure, although the precise combination can vary. The revised EU ETS Directive 2009/29/EC,[3] adopted as part of the climate and energy legislative

[3]As briefly discussed in Section 5, the European Parliament and the Council

package in April 2009, details how to identify the sectors at a risk of carbon leakage (European Commission, 2009a). Indeed, such sectors are identified based on their cost impact and trade-exposure, but the Directive relates to the thresholds different from those aforementioned empirical studies. In addition, trade intensity is listed as a separate criterion in the EU Directive. According to the Directive, a sector would be deemed to be exposed to a significant risk of carbon leakage if one of the following three conditions is met:

- 5% cost impact and 10% trade intensity: (direct and indirect carbon costs) / gross value added $\geq$ 5% and (exports and imports) / (EU turnover and imports) $\geq$ 10%
- 30% cost impact: (direct and indirect carbon costs) / gross value added $\geq$ 30% or
- 30% trade intensity: (exports and imports) / (EU turnover and imports) $\geq$ 30%

In calculating the direct cost of the required allowances and the indirect cost from higher electricity prices resulting from the implementation of the Directive, the European Commission has assumed an allowance price of € 30/t CO_2, 75% auctioning, and the average emission factor for electricity generation in the EU of 0.465 t CO_2 per MWh (European Commission, 2009b). As a general rule, the trade data for 2004-2006 and the CO_2 costs for 2005-2006 have been taken in these determinations. The quantitative analysis is carried out at the four-digit classification level of the NACE (Classification of Industries Established in the European Communities) (258 sectors in total), unless unavailable. The sectors identified to be at a significant risk of carbon leakage then qualify for receiving 100% of allowances for free along benchmarks for each of the sectors identified. No free allocation shall be given to electricity generators, meaning that even those sectors receiving all allowances for free will have to bear indirect cost from

(2008) amended in February 2018 the EU ETS Directive. While the reform of the EU ETS relates mainly to the period after 2020, the so-called fourth phase (2021-30) of the EU ETS, it involves issues, such as the reduction of the current surplus of allowances in the EU carbon market in the third phase.

electricity price increase as a consequence of CO_2 price pass through in the electricity sector. For other sectors, the Directive suggests that 80% of allowances are handed out for free in the initial year (2013) of the third phase, with the share of free allowances declining to 30% by 2020, the end year of the phase. Such free allocations are based on the *ex ante* benchmarks that are set at the average performance level of the 10% most efficient installations in a given sector or subsector in the EU in the years 2007-2008 (European Commission, 2009a). This suggests that such benchmarks represent a challenge for some installations because they are set at the level of the best performers, but they are achievable by definition because they are derived from real practice in recent years. Moreover, benchmark values will be updated during the fourth phase (2021-30) to reflect technological progress in the different sectors or subsectors (European Parliament and the Council, 2008).

These criteria sharply contrast with the original Commission proposal. The European Commission originally proposes only the first criterion. Following an intensive lobby from industry, the second and third criteria are included in the revised Directive. Moreover, most of the identified sectors are attributed to the third criterion. Of 146 sectors identified to be exposed to a significant risk of carbon leakage at the NACE 4-digit level, 117 sectors are due to the third criterion, showing a high trade intensity. Other 27 sectors have both significant CO_2 cost and trade intensity. Two sectors qualify due to the significant CO_2 cost criterion alone: cement (45.5%) and lime (65.2%) (European Commission, 2009b).

In H.R. 2998 (U.S. House of Representative, 2009), sectors eligible for allowance rebates are determined at the six-digit classification level in Codes 31-33 of the North American Industrial Classification System. A sector is determined to be eligible for allowance rebates if it meets with either of the following conditions.

- 5% energy or greenhouse gas intensity and 15% trade intensity: (energy or greenhouse gas costs) / the shipments $\geq$ 5% and (imports and exports) / (the value of its shipments and imports) $\geq$ 15% or
- 20% energy or greenhouse gas intensity: (energy or greenhouse gas costs) / (the value of its shipments) $\geq$ 20%

Clearly, the U.S. House bill differs from the aforementioned EU Directive, based on a minimum level of energy or greenhouse gas intensity or a combination of a minimum level of energy or greenhouse gas intensity and a minimum level of trade exposure to identify the sectors at a risk of carbon leakage. A level of trade intensity alone is not a criterion.

This bill specifies the annual average data for 2004-2006 to be used in these determinations, unless unavailable. However, the bill provides that U.S. Environmental Protection Agency shall determine additional sectors to be eligible if they 1) meet the energy or greenhouse gas intensity criteria at the time the rule is promulgated and 2) meet trade intensity criteria based on post-2006 data.

During deliberations of the bill, the Energy-Intensive Manufacturers' Working Group on Greenhouse Gas Regulation compiled a list of the sectors that meet with the eligibility criteria. 47 sectors are identified to be eligible for allowance rebates (McMackin, 2009). This list is identical to the list compiled by the U.S. EPA (2009), with paperboard and beet sugar not included in the EPA list due to differences in data sources. This number is considerably less than the 146 sectors potentially covered by the EU Directive. The main reason for this difference is that not all EU criteria have taken into consideration the primary factors influencing a company's ability to compete under a carbon constraint: 1) the greenhouse gas intensity of its products, 2) its ability to pass on any increased costs to consumers without losing market share or profitability; and 3) its ability to mitigate carbon emissions (Parker and Grimmett, 2009).

The 30% trade intensity criterion in the EU Directive alone has added 117 sectors to the list (146 sectors identified in total). They include everything from the manufacturing of wines to textiles. These sectors are trade-exposed, but are not considered to be carbon intensive. This raises the issue of which criterion, carbon intensity or trade intensity, provides a good basis to identify the sectors at risk of carbon leakage. Sato *et al.* (2015) suggest that carbon intensity measured as cost increase relative to gross value added provides a robust metric across countries and across time, to compare potential carbon cost impacts between sectors. Based on a firm-level analysis of the EU ETS, Martin *et al.* (2014)

find that carbon-based criterion is a good indicator to measure the risk of downsizing whereas trade intensity criterion is not. Because trade intensity criterion is not related to carbon and thus allowance prices, these sectors identified based on the criterion will receive all allowances for free, even if the allowance prices are at the very low levels as they are now. Clearly, the European Commission has been extraordinarily generous in its identification of carbon leakage exposed sectors.

In screening sectors, both the EU and the U.S. have used a single approach across all sectors for free allowance allocations. Dröge *et al.* (2009) suggest that screening needs to take into consideration the cost structures and the investment options. That means that not only both direct and indirect costs, but also capital intensity of production, new investment need and the diversification of products should be considered. Moreover, screening should be linked with the best measures to address carbon leakage for that sector. Along this line, for sectors with high direct costs, if their production is capital intensive, reducing their carbon emissions needs new investment. Just like blast furnaces for steel production, the substitution of this process to cut carbon emissions needs new investment. A direct support could help steel firms not relocate elsewhere. If free allowances are applied, a new entrants reserve, which is set up to 5% of EU wide allowances during the third phase of the EU ETS from 2013 to 2020, could be used as an incentive for investors. On the other hand, if production is not capital intensive, and running a plant below full capacity is possible like clinker production, adjustments at the border work better for a homogenous product like clinker than for heterogeneous products (see Figure 3.1).

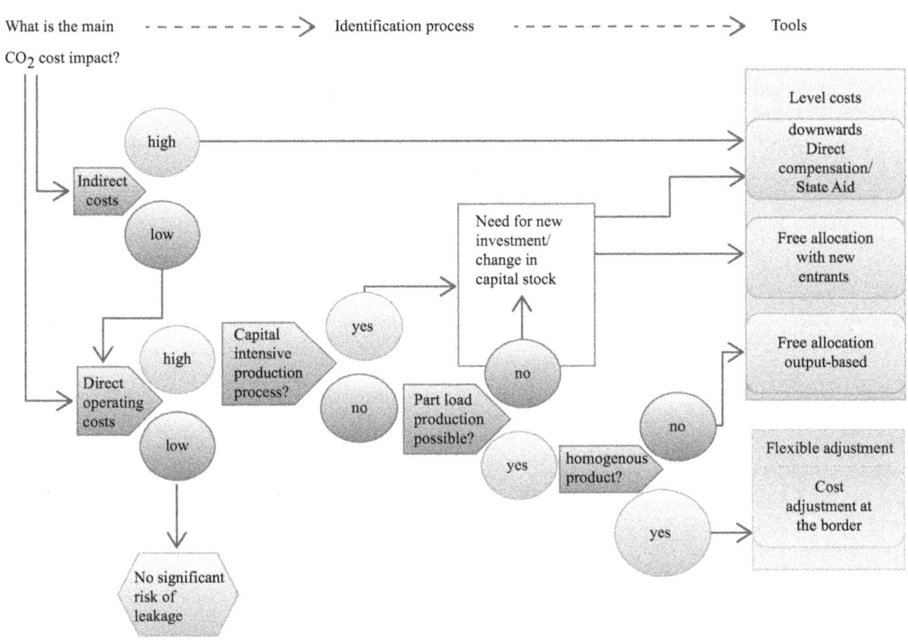

Figure 3.1: Screening of leakage potential
Source: Dröge *et al.* (2009).

4

How large is the magnitude of leakage impacts?

Quelle est l'ampleur de l'impact des fuites ?

We will first discuss *ex ante* estimates of potential carbon leakage rates. We then examine the environmental tax reforms and greenhouse gas emissions trading schemes that have been implemented to provide an *ex post* assessment of their competiveness and carbon leakage impacts to see a difference, if any, with those *ex ante* studies. Finally, we will explain why they differ from each other.

4.1 *ex ante* studies at the aggregated and sectoral levels

There have been a large body of *ex ante* studies that have attempted to quantify leakage impacts of unilateral or uneven climate policies. Work in this area can be traced back to Pezzey (1992), Rutherford (1992), and Felder and Rutherford (1993). These and subsequent studies provide wide-ranging estimates of carbon leakages, depending on their assumptions on returns of scale, behaviors in the energy-intensive industries, price elasticity of demand, elasticities of trade substitution between domestic and imported products, supply elasticities of fossil fuels, transport costs, carbon cost pass-through capacity, and other factors (Burniaux and Martins, 2000; Mathiesen and Maestad, 2004; Babiker, 2005; Babiker and Rutherford, 2005; Gerlagh and Kuik, 2007;

IPCC, 2007; Reinaud, 2008b; Carbone and Rutherford, 2009; Böhringer *et al.*, 2010, 2014 and 2017a,2017b; Monjon and Quirion, 2011). For example, Burniaux and Martins (2000) show that carbon leakages are highly sensitive to the assumed values of fossil fuel supply elasticities. Saito (2004) shows that the existing uncertainties in Armington (1969) elasticities whose values reflect how ease substitution between domestic and imported goods is, might not only affect the magnitude of policy effects, but possibly even the sign (positive or negative). When Armington elasticities are higher, foreign and domestic varieties of traded goods are close substitutes. This might result in stronger trade and leakage effects of climate policy. The carbon leakages via the international fossil fuel channel decrease with the supply elasticity of fossil fuels (Gerlagh and Kuik, 2007), and the carbon leakages linked to the competitiveness channel increase with the substitutability between domestic and imported products (Bernard and Vielle, 2009; Carbone and Rutherford, 2009; Böhringer *et al.*, 2017a). Moreover, a carbon leakage depends on the size of carbon-abating coalition. The more countries take carbon abatement, the less is the carbon leakage (Reinaud, 2008b; Carbone and Rutherford, 2009; Böhringer *et al.*, 2014; Branger and Quirion, 2014).

A literature review by IPCC (2007) suggests that the estimates of carbon leakages from implementing the Kyoto Protocol are generally in the range of 5-20% by 2010, very much in line with the range of 2-21% from the survey of Gerlagh and Kuik (2007) and 2-23% from the survey of Dröge *et al.* (2009). Branger and Quirion (2014) analyze 310 estimates of carbon leakage and output loss in energy-intensive and trade-exposed sectors from 25 studies dating from 2004 to 2012 and suggest that the carbon leakage ratio ranges from 5% to 25%. That is, for every four to twenty tons of cuts in CO_2 emissions in an abating country, one additional ton will be emitted outside this abating country.

One estimate slightly below the lower bound is from Mattoo *et al.* (2009). This World Bank study finds that the carbon leakage rate is only 1% when high income countries unilaterally cut their emissions in 2020 by 17% relative to their emissions levels in 2005. One reason for this rather small leakage effect is that while the unilateral emissions reductions by high income countries increase exports of carbon-intensive products from countries like Brazil, China and India, what matters for emissions is the impact on these countries' overall output and its composition. Since exports are a small proportion of output, the increase in output of carbon-intensive sectors in these three countries is only about 1-2%. Another reason is that this expansion of these energy-intensive sectors pulls resources out of other sectors, which has an offsetting effect on emissions.

One notable outlier significantly goes beyond the upper bound in the three surveys, reporting a leakage rate of 130% (Babiker, 2005). Leakage rates above 100% mean that carbon reduction in one region is more than offset by emission increase outside this region, thus leading to more global emissions rather than less. This is not impossible if implementing a strict climate policy imposes additional costs on energy- and carbon-intensive industries being unable to pass on these costs to their consumers. This will drive demand and production of products and the resulting emissions to the shifts from carbon-constrained countries to countries without similar carbon constraints. Given that emissions intensities in the abating countries are typically lower than in the non-abating countries, such a shift would lead to more emissions for producing per unit of like product. Another and probably most likely explanation is that, as discussed in Section 2, a large part of carbon leakage occurs through the international fossil fuel channel, which is unaffected by anti-leakage policies. However, having a close look at the analysis, it is found that this outlier rests on two assumptions: the homogeneity of goods and increasing returns to scale, which differ from the usual assumptions of products differentiation and constant returns to scale under the computable general equilibrium modeling. Babiker (2005) shows that when perfect homogeneity is assumed, increasing returns to scale assumption increases a carbon leakage to 130% from 60% under the constant returns to scale assumption. By contrast, in the case of the differentiated trade in products, increasing returns to scale assumption only produces a carbon leakage rate of about 25%. Clearly, this homogeneity assumption alone, no matter of returns of scale, has a significant effect on a carbon leakage. Carbone and Rutherford (2009) support this finding, suggesting that carbon leakage becomes a considerable problem when traded goods from different regions are close substitutes under the assumption of homogeneity of goods (homogenous trade).

The aforementioned carbon leakages are estimated at the aggregated level. But because energy-intensive industries are major carbon emitters, it should come as no surprise that climate policies have targeted at these industries, such as steel, cement and aluminum. Indeed, the EU ETS targets cement and steel, the two largest CO_2 emitters among European manufacturing sectors, with each receiving 10% and 9% of the allowances allocations in the EU ETS, respectively (Kettner *et al.*, 2013). These sectors are relatively open to international trade, ease to relocate and have some degree of product and process uniformity—consumers tend to be indifferent to the origins of these products, provided that they are less expensive. Therefore, these energy intensive sectors are exposed to a great risk of carbon leakage. The aggregated numbers hide very different carbon leakages at the sector level. Thus, assessing carbon leakage needs to focus on more detailed sectors (Monjon and Quirion, 2011).

An increasing number of sectoral (bottom-up) studies have taken into consideration of the specific energy-intensive industries in order to investigate the effects on production and location decisions by firms, thus determining how large leakage rates are for these sectors exposed to global competition. Given that higher leakage rates would be expected for more trade-intensive sectors and that steel is much more traded on an international level, let us take the iron and steel sector as an example.[4] Gielen and Moriguchi (2002) develop STeel Environmental strategy Assessment Program (STEAP), a large scale partial equilibrium model that covers the life cycle analysis of many different technology options in the iron and steel sector. They find that if Japan and Europe alone introduced such a carbon tax, their CO_2 emissions would indeed decline, but that lower production in these regions would be offset by increased production and emissions elsewhere. They show that a tax of US$10-42/t CO_2 would generate a carbon leakage rate of 35-75% by 2020 in the iron and steel sector. According to the STEAP model, marginal tax increases lead to much higher increases in leakage at low tax levels (US$10–50/t

[4]Steel has a much higher value-added per tonne than cement (roughly ten times more) and is more widely traded (Branger and Quirion, 2013). In 2011, 31% of finished steel products were internationally traded (WSA, 2012), while only 3.9% of cement was traded internationally in 2012 (ICR, 2013).

CO_2) than at high tax levels (around US\$100/t CO_2 and beyond). This non-linear behavior indicates a strong sensitivity of leakage to small tax increases compared to the status quo (Oikonomou *et al.*, 2006). A study by the OECD (2003) also concludes that an OECD-wide carbon tax of US\$ 25/t CO_2 in the steel sector would lead to a carbon leakage rate of 45% in the steel sector. If the implementation of this carbon tax at the same level is not coordinated across the OECD countries, rather unilaterally, this leakage rate then would, on average, increase to 60%. Using a global partial equilibrium model of the steel industry, Mathiesen and Maestad (2004) simulate potential effects on world CO_2 emissions of the introduction of a US\$ 25/t CO_2 tax in the steel sector in Annex I countries. That would lead to a carbon leakage rate of 26%. If the substitution possibilities between different types of steel and between inputs to steel production were not included, this estimated leakage rate in the steel industry would double, increasing to 53%. Quirion (2009) applies a small partial equilibrium model that uses marginal abatement costs from the large POLES (Prospective Outlook for the Long term Energy System) model. Their results suggest a leakage rate of 45% in the steel sector for a 15% reduction in emissions in the EU in 2015 relative to 2005 levels.

Clearly, the estimated carbon leakage rates in the iron and steel sector differ greatly with the models and assumptions used in these sectoral studies. However, none of these simulations at sectoral levels have indicated a leakage rate near 100%. This implies that it is highly unlikely that carbon leakage would offset entirely carbon abatement in a given industry. This is different from the aforementioned studies at the aggregated level, which indicate that carbon leakage rate could be above 100% under the specific assumptions.

4.2 *ex post* studies on ETRs and ETS

The OECD countries, in particular the EU member states, have implemented environmental tax reforms and greenhouse gas emissions trading schemes. These experiments provide a basis for an *ex post* assessment of their competitiveness and carbon leakage impacts to see a difference, if any, with those *ex ante* studies. A body of *ex post* studies, although there are a few, have examined their impacts after implementing these market-based environmental instruments and have compared with the references before the introduction of these instruments (Agnolucci, 2009; Andersen *et al.*, 2007; Andersen and Ekins, 2009; Barker *et al.*, 2009; Ellerman and Buchner, 2007 and 2008; Enevoldsen *et al.*, 2007; Convery *et al.*, 2008; Reinaud, 2008b; Ellerman *et al.*, 2010; European Commission, 2012c; Grubb, 2012; Kettner *et al.*, 2012).

Let us start with environmental tax reforms. Staring in Finland and Sweden in 1990, a number of European countries have implemented environmental tax reforms (ETR), shifting the burden of taxes from 'goods' (e.g., labor and capital) to 'bads' (e.g., carbon emissions). Since environmental taxes serve to correct market failures, such a tax shift could improve short-term economic performance, while delivering a long-term environmental dividend, thus leading to the so-called double dividend (Pearce, 1991). The COMETR (COMpetitiveness effects of Environmental Tax Reforms) project (Andersen *et al.*, 2007; Andersen and Ekins, 2009) is a comprehensive attempt to provide an *ex-post* assessment of the environmental and economic effects of the ETRs implemented in seven EU member states (Denmark, Finland, Germany, the Netherlands, Slovenia, Sweden, and the UK) at the macro level as well as at a sector level. Generally speaking, the Nordic countries generally provide the firmest basis for an *ex-post* assessment because they have time series of more than a decade since the introduction of ETRs, in comparison with Germany and UK where carbon-energy taxes were introduced only in 1998 and 2000. In order to disentangle the specific impact of the ETRs, two scenarios were developed with the E3ME model, a comprehensive time-series estimated macro-economic model of eleven different fuels and more than 40 economic sectors within the EU. One scenario is labeled as the 'baseline case'. This scenario was calibrated closely to the observed outcomes over the historical period from 1994-2003 for all seven EU member states with ETRs and the rest of the EU, using historical data that include the effects of ETR implementation. Another scenario is termed as the 'counterfactual reference case'. This reference scenario basically involved a projection of 'what if' there had been no tax shifting as a result of the ETRs, keeping everything else in the model constant but including current and expected developments in the EU economy (Barker *et al.*, 2009). This counterfactual exercise allows to isolate the effects of the ETRs.

Barker *et al.* (2009) suggest that the double dividend theory proves true. All six EU member states experience reduced CO_2 emissions but none of them experience negative GDP. No reduction is identified in Slovenia, which in fact mainly relabeled its pre-existing mineral oils tax into a CO2-tax. By 2004, the ETRs had, on average, led to reductions in greenhouse gas emissions of 3.1% for the six member countries. The largest reduction in emissions occurs in Finland and Sweden, as a result of their highest tax rates. By contrast, the ETR is not particularly efficient in reducing emissions in Germany as coal was not included. Depending on how the revenues from environmental taxes are recycled, five of the ETR countries have an increase in GDP. In Sweden, the effects take slightly longer to come through, as the very large increase in household electricity taxes depresses real incomes in the short run. Finland experiences a short-term GDP boost from the effects of the taxes on fuel demand, because a reduction in the demand for imported fuel improves its trade balance. In the UK there is the no discernable effects on GDP. This can be attributed to the fact that the scale of tax rates (i.e., climate change levy on fossil fuels) levied and the revenues generated and recycled back to the economy has been rather modest (Andersen *et al.*, 2007; Andersen and Ekins, 2009).

Up to now, an assessment is made on the ETR countries. It would be interesting to see how the ETR countries perform relative to countries that have not implemented the ETR, i.e., non-ETR countries. Barker *et al.* (2009) and Ekins (2012) show that the ETR countries experience slightly faster GDP growth than they had without the ETR, and the non-ETR countries in the EU show practically no change. This suggests that, far from damaging the competitiveness of ETR countries compared to the non-ETR countries, the ETR countries benefit economically, as well as environmentally with CO_2 emissions falling in the ETR countries collectively, from the ETR policy. Moreover, the ETRs have almost no effect on the level of CO_2 emissions in the non-ETR countries (Figure 4.1). This suggests that there was no carbon leakage from the ETR countries collectively to the non-ETR countries in the EU.[5]

[5]However, an important caveat exists as the E3ME model from which the results

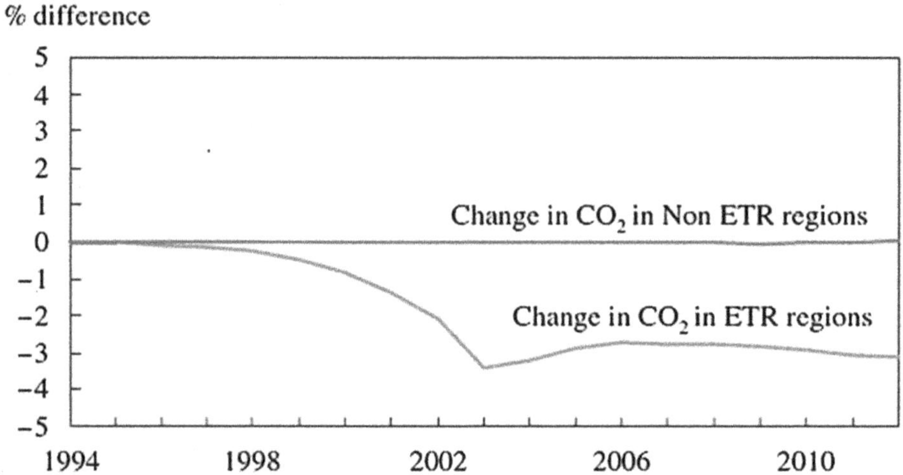

Figure 4.1: Changes in CO_2 emissions in the ETR and non-ETR countries (% difference between the base case and the counterfactual reference case)
Source: Speck (2007).

Ekins and Salmons (2007) and Miltner and Salmons (2009) have analyzed *ex post* effects of the ETR on the following eight sectors at NACE 3-digit level in the seven EU member states: Meat and meat products (15.1), paper products (21.2), basic chemicals (24.1), pharmaceuticals (24.4), glass and glass products (26.1), cement, lime and plaster (26.5), basic ferrous metals (27.1-3), and basic non-ferrous metals (27.4). They examine the changes in competitiveness indicators (the share of global production, import intensity and export intensity) for each country and each sector according to the percentage change in unit cost of production. Table 4.1 lists the results for the 56 combinations of eight sectors and seven countries.

The ETRs have not been significant in terms of their impact on unit costs of production (below 1% in 50 cases). Moreover, losses in competitiveness occur only in countries/sectors where the impact of the ETRs on unit costs has been no larger than 1%. There is no case of a decrease (or increase) in competitiveness where the impact of the ETRs was above 1%. Out of 56 cases considered, the results do not

are derived is not a world model, implying that the abovementioned result is limited to the EU member states.

Table 4.1: Sectoral competitiveness analysis in the ETR countries

% change in unit costs	Gain in competitiveness	No change in competitiveness	Loss in competitiveness	Total
Less than 1%	2	39	9	50
1%-5%	0	5	0	5
More than 5%	0	1	0	1
Total	2	45	9	56

Source: Ekins and Salmons (2007).

support for a change in competitiveness in 45 cases (namely, 80% in 56 cases) relative to their competitors. Put another way, competitiveness in 45 cases does not change with or without the ETRs relative to their competitors. Only in nine cases (16%) do the trends in the indicators suggest a loss in competitiveness. These occur in the meat and meat products (15.1), paper products (21.2), basic ferrous metals (27.1-3), and basic non-ferrous metals (27.4) in the UK; the paper products (21.2), pharmaceuticals (24.4), and glass and glass products (26.1) in Germany; the pharmaceuticals (24.4) in Finland; and the basic non-ferrous metals (27.4) in the Netherlands. In 4% of 56 cases, there is actually an improvement in competitiveness in the Danish pharmaceuticals industry and the Dutch meat processing sector (Miltner and Salmons, 2009). However, some qualifications need to be added to this finding. A comparison of ETRs applied to the aforementioned eight sectors in Denmark, Finland, Germany and the UK reveals that, although with exceptions, sectors of low energy intensities tend to have high or medium tax rates, while sectors of high energy intensities have low or medium tax rates (Miltner and Salmons, 2009). In absolute terms, unilateral energy/carbon taxes in these countries have been modest, and for the most energy-intensive sectors, they are very low indeed or to some extent symbolic (Andersen *et al.*, 2007; Andersen and Ekins, 2009; Miltner and Salmons, 2009).

Now let us turn into emission trading experiments. Since January 2005, the EU put into operation the world's largest multi-country, multi-sector carbon dioxide emissions trading scheme (ETS). The EU ETS has provided a large-scale experiment to identify the magnitude of competitiveness and carbon leakage. Analysis of cement, iron and steel, aluminum and refinery sectors does not reveal carbon leakage for these trade-exposed carbon-intensive sectors during the first phase (2005-2007) of the EU ETS. There is also no evidence that the EU ETS in place has triggered changes in trade flows and production patterns in these sectors during this phase (Reinaud, 2008b). Based on various econometric methods, Branger and Quirion (2013) perform the two estimations of the carbon price on the net imports of cement and steel: one without the carbon price over the time period 1999-2012, and another with the carbon price over the time period 2004-2012. They find that the carbon price is not statistically significant: the carbon price has no impact on the net imports of cement and steel during the first and second phases of the EU ETS. Rather, net imports of cement and steel have been driven by domestic and foreign demand. On this basis, they conclude that, at least in the cement and steel sectors, the EU ETS has not generated competitiveness losses and carbon leakage either in its second phase.

4.3 Why *ex post* results differ from *ex ante* projections?

The aforementioned *ex post* results from this world's unprecedented, grand climate policy experiment are in sharp contrast with *ex ante* studies that have projected much higher carbon leakage effects. There are two sets of explanations for this difference. The first set of explanations are related to the assumptions and data used by *ex ante* studies that do not reflect the reality of the EU ETS. Another set of explanations for *ex ante* overestimate of the net leakage effects concern the shortcomings of models themselves that omit the potential positive impacts of climate policy.

Let us start with the first set of explanations for this difference. First and most importantly, this is due to the free allocation of allowances to the entities covered under the EU ETS. The EU ETS Directive allowed member states to auction up to 5% of allowance allocations in the first phase 2005-07, rising to 10% in the second phase 2008-12. In the first phase, only four of 25 member states used auctions at all, and only Denmark auctioned the full 5% (Ellerman and Buchner, 2007). As a result, auctioned allowances accounted for 0.13% of the total allocation (Convery *et al.*, 2008). In the first four years of the second trading phase, auctioned volumes increased, but only accounted for about 4% of total allowances issued (European Commission, 2012c).[6] Moreover, in most cases allowances are over-allocated to the industry sector, with electricity and heat generation being the only sector that exhibits a shortage of allowances in all trading years from 2005 to 2010 (Kettner *et al.*, 2012). Put another way, not only receiving allowances for free, the industry sector even gets more than its needs. Analysis of the Community Independent Transaction Log data suggests the number of the allocated allowances 7% above the level of reported emissions for cement and lime installations in the first phase (Ellerman and Buchner, 2008). This surplus of allowances reaches about 17.5% for the iron and steel sector (Convery *et al.*, 2008) and about 20% for the pulp and paper sector. For the EU 23 (the old EU 15 plus 8 new member states from Eastern Europe) as a whole, verified emissions are 3.1% less than their baseline emissions (Ellerman and Buchner, 2008) (see Box 1). The over-allocation of allowances has led to significant drop in allowance prices. When Belgian, Czech, Dutch, and French verified emissions were released, it was shown that their emissions were much lower than that allowed. This surplus of allowances let to allowance prices dropping by 55% in only three days. The price dropped from a record high of € 31/t CO_2 in mid-April to only € 13.5/t CO_2 by 27 April 2006 (Zhang, 2006). In the second trading phase the European Commission took more tough position in limiting the Member States' generous allocations in their initial National Allocation Plans. Most proposed national caps had to be adjusted downwards by the Member States after the European Commission's review process. This led to substantial changes in allocation discrepancies for several Member States, in particular the Baltic States (Capoor and Abrosi, 2008). In the first trading phase the New Member States generally exhibited higher net long positions than the EU-15 and only one of the New Member States (Slovenia) but five EU-15 countries were in a net short position. In the first three years of the second trading phase, these

[6]In the second phase, eight member states used auctions, and about 3% of the allowances were auctioned (ICAP, 2018). The European Commission estimates that 57% of the total amount of allowances will be auctioned during the third phase 2013-2020. From 2021 onwards, the share of allowances to be auctioned shall be 57% (European Parliament and the Council, 2008). See the auctioning on the European Commission's web site at https://ec.europa.eu/clima/policies/ets/auctioning_en.

regional differences largely disappeared as a result of the Commission's intervention in National Allocation Plans (Kettner *et al.*, 2012). As a result, the overall EU cap for 2008 was binding, with verified emissions exceeding allocated allowances on aggregate by 5.3%. For 2009, because of the exogenous shock of the economic crisis to the installations in the trading system, however, allocation again exceeded verified emissions by 8.8%. For 2010 emissions moderately increased, but allocated allowances still exceeded verified emissions. This surplus gap become even bigger in 2011 because a decline in verified emissions as a result of continuous economic growth stagnation. Overall, verified emissions exceeding allocated allowances on aggregate by 5.2% during the period 2008-11 (European Commission, 2012c). Despite the European Commission's tough position in limiting the Member States' initial generous allocations for the second phase, this period will end up with a huge volume of unused emissions allowances that can be banked forward into the third phase. Estimates of this surplus differ. The Commission's own estimate suggests that this surplus potentially represents the equivalent of 2.4 billion tons of allowances by 2020 (European Commission, 2012a), while other estimate puts this surplus at 3.1 billion tons of allowances (Morris, 2012). For electricity-intensive sectors like aluminum, the still functioning long-term electricity contracts limit their actual exposure to rising electricity prices. Only 18% of the EU capacity operated without long-term electricity contracts. While the details of these contracts are unknown, cost estimates suggest lower average electricity costs in these cases than for smelters in Germany and the Netherlands acquiring electricity on the market. The two countries see that their power prices doubled during the period 1999-2006, relative to small changes in other EU member states (Reinaud, 2008a; Convery *et al.*, 2008). Furthermore, booming demand for aluminum and other commodities has kept their prices high, thus making all manufacturers profitable. Given that any impacts of the carbon prices on industry are likely to be felt more strongly when markets are less favorable, high commodity prices during the period 2005-07 make it difficult to observe any effects of carbon prices on these sectors' profitability. Product specifications that vary from country to country, meanwhile, help to protect European refiners from foreign competition (*The Economist*, 2008a; Reinaud, 2008b; Convery *et al.*, 2008).

Box 1 Over-allocation or abatement in the pilot phase of the EU ETS?

A long position has been widely interpreted as evidence of over-allocation of allowances. However, it is unsettled whether it is a result of over-allocation, abatement, or both. A surplus of allowances could be because of an over-estimate of the level of CO_2 emissions and the consequent demand for allowances as a result of rising real output, the adverse weather in 2005 and the higher prices of natural gases relative to coal, which would have driven increased use of coal and thus lead to increased emissions and demand for the allowances. Another possibility for a surplus of allowances could be an under-estimate of the amount of abatement that would occur as the affected facilities incorporated carbon prices into their production decisions. Ellerman and Buchner (2008) argue that this under-estimate of the amount of abatement is a distinct possibility, given that a significant price was paid for CO_2 emissions during the period 2005-06, which would have the effect of abating CO_2 emissions as firms adjust to this new reality (Convery *et al.*, 2008). Their preliminary estimate shows that the EU ETS had reduced at least 50 million tons of CO_2 emissions annually. Therefore they conclude that only a relatively small proportion of the observed long position of nearly 60 million tons for the EU as a whole in 2005-06 can be attributed to over-allocation. Anderson and di Maria (2011) also support the contribution of abatement activities by analyzing abatement and over-allocation in the ETS pilot phase based on a dynamic panel data model. Using data on historical CO_2 emissions, economic activity, electricity prices and climate factors they find in 2005, 2006 and 2007 a net abatement of 84.2, 61.7 and 27.6 million tons of CO_2 emissions, respectively. The total abatement of 173.5 million tons of CO_2 emissions in the pilot phase was 2.8% of the estimated baseline emissions. Over the pilot phase, they find that the EU25wide cap was set only 27.9 million tons of CO_2 emissions lower than the estimated baseline emissions across the entire trading community, or 0.45% below the EU25 baseline emissions. Based on a review of published estimates, Grubb (2012) suggests an average yearly abatement since the establishment of the EU ETS in the range 30-70 million tons of CO_2 emissions, roughly 2-4% of the total emissions covered.

Now let us move to the second set of explanations for the difference between *ex ante* projections and *ex post* results. Copeland and Taylor (2005) have developed a theoretical model that examines a carbon unconstrained country's best response to a carbon emissions reduction in the emissions constrained country. Taking into account a free rider effect, carbon leakage (a substitution effect), and an income effect, they show that in an open trading world, unilateral emission reductions by a set of rich Northern countries can create self-interested emission reductions by the unconstrained poor Southern countries. The rational is that since the unconstrained country is a dirty good exporter, its emissions will tend to rise via free rider effects and substitution effects in production, but will tend to fall via substitution and income effects in the demand for environmental quality. Overall, the increase in emissions in the unconstrained country may be small or even zero.

High estimate of carbon leakage is because modeling studies have typically regarded the technology in each country as given. They do not take into account induced technological innovation and diffusion brought about by the implementation of climate policies. In the short to medium term with production capacities largely given, this does not seem like a major restriction. However, in the longer term, considerable technological developments could take place in the sector if carbon taxes or tradable permits were to be applied. Indeed, there is evidence that abatement technologies are endogenous, and thus technology development is affected by energy policies or environmental policies (Newell *et al.*, 1999; Jaffe *et al.*, 2002; Löschel, 2002; Popp, 2002). Grubb *et al.* (2002) has estimated the potential impact of international spillovers due to mitigation actions by the industrialized countries on greenhouse gas emissions in the developing countries. They show that spillovers from Annex I mitigation actions, via induced technological change, could have substantial effects on sustainable development, with emission intensities of developing countries at a fraction of what they would be otherwise.

Golombek and Hoel (2004) demonstrate that with endogenous technologies and technology diffusion between countries, reduced emissions in some constrained countries might also reduce emissions in other unconstrained developing countries. Bosetti *et al.* (2008) examine effects of international energy research and development (R&D) spillovers on the costs of stabilizing atmospheric CO_2 concentrations at 450 ppmv at the end of the 21st century by analyzing a policy-mix in which a climate policy based on a global permit market is coupled with a technology policy based on transfers. They show that global stabilization costs would be much lower when resources are used to increase the absorption capacity in developing countries, instead of a simpler, non-targeted, lump-sum transfer of the same amount of resources from high income countries to developing countries. This is because existing barriers have prevented developing countries from absorbing international knowledge spillovers.

In empirical energy and climate modeling studies on the spillover effects, induced technological change can be implemented via R&D and learning-by-doing (LBD) channels (Sijm *et al.*, 2004; IPCC, 2007; Gillingham *et al.*, 2008; Pizer and Popp, 2008). Through R&D channel, the introduction of climate policies increases the market for carbon-mitigation technologies and thus creates an incentive for increased R&D investments in these technologies. In modeling terms, these increased investments lead to an increase of the knowledge capital stock. Models used to endogenize the effect of R&D on technology and mitigation costs have generally included a variable representing R&D that influences economic behavior through one of three routes: a direct impact on the level of emissions intensity, a reduction in the mitigation cost function, or productivity gains in sectoral production functions (with sector-specific R&D stocks in multi-sectoral models) (Pizer and Popp, 2008). Depending on approaches to modeling spillovers and opportunity costs, studies differ regarding effects from endogenizing R&D, with relatively small effects found from Nordhaus (2002) and much larger effects from Buonanno *et al.* (2003).

LBD channel means that climate policies encourage primarily the adoption of carbon-mitigation technologies, resulting in declining costs of these technologies due to the accumulation of knowledge and experience as the installed capacity of these technologies expands. In modeling terms, this process of technological change is expressed by a learning curve that relates the costs of a technology to its cumulative installed capacity Sijm *et al.* (2004). Söderholm and Sundqvist (2007), and Söderholm and Klaassen (2007) have extended this basic presentation to estimate "two-factor" learning curve that models cost reductions of a technology as a function of both its cumulative capacity (LBD) and R&D (learning by searching). Söderholm and Sundqvist (2007) find that learning by searching, rather than learning by doing, contributes more to cost reductions. The literature on learning curve shows that a number of new low-carbon technologies for power generation have indeed experienced significant cost reduction as they are progressively deployed (McDonald and Schrattenholzer, 2001; Junginger, 2005; IPCC, 2007; Junginger *et al.*, 2010). Given that learning tends to be thought of as a technology-specific phenomenon, disaggregated, technology-oriented bottom-up models for energy and climate policy analysis have widely incorporated learning for each technology. Sijm *et al.* (2004) find that most top-down modeling studies omit the spillover effect or show it playing a minor role. They point out that while the potential beneficial effect of technology transfer to developing countries arising from technological development brought about by Annex I action may be substantial for energy-intensive industries, it has so far not been quantified in a reliable manner. Even in a world of pricing CO_2 emissions, there is always likelihood that net spillover effects are positive, given the unexploited no-regret potentials and the technology and know-how transfer by foreign trade and educational impulses from Annex I countries to Non-Annex I countries (Sijm *et al.*, 2004). In any case, they conclude that, in practice, carbon leakage is unlikely to be substantial because transport costs, local market conditions, product variety and incomplete information all favor local production (Sijm *et al.*, 2004; IPCC, 2007).

Carbon leakage may also be overstated or even the size of carbon leakage might be misstated because abatement resource effect is not considered. The so-called abatement resource effect refers to that the taxed sector draws resources away from the other sector or country, which reduces their output and emissions. Baylis *et al.* (2014) show that if consumers can shift their purchases easily, leakage may be overstated in models that do not allow for substitution in production. If consumer flexibility is low compared to producers' ability to abate pollution by use of other resources, overall leakage may be negative. Zhang and Zhang (2017) further examine abatement resource effect on carbon leakage both theoretically and empirically, decomposing the abatement resource effect into the consumption substitution effect and the multiplier effect and deriving closed-form solutions for these leakage effects. Based on a carbon price on the electricity generation sector in China, the decomposition analysis suggests that of the four effects (namely, the scale effect, the consumption substitution effect, the multiplier effect, and the production substitution effect) identified, such leakage is mainly through the production substitution effect and the multiplier effect (Zhang and Zhang, 2017).

Regardless of these mechanisms and considerations, how baselines are simply defined in *ex ante* studies matter. The abating countries expect some anticipated technological changes induced by existing polices even in the baselines. This is particular true for R&D that improves energy efficiency, which is encouraged by higher energy prices that may or may not be the result of environmental policy (Popp, 2002). Thus, appropriate baselines for any model that is calibrated to current data trends should incorporate anticipated changes not directly as a result of climate policy, rather by the existing policies. If baselines simply assume no policy, that would set high baseline for the abating countries relative to baselines for non-abating countries than if baselines incorporate the existing policies. That will lead to higher *ex ante* carbon leakage rates.

There is some evidence for a first-mover advantage for specific renewable energy technologies, e.g., wind power and solar photovoltaics (PV) for Germany. The development of solar PV in Germany shows that having an appropriated policy does make a difference. While Germany has unfavorable solar radiation conditions compared with its Southern European countries, thanks to feed-in tariffs, it leads the world in both accumulated installations and new photovoltaic additions. Its cumulative PV installations through 2009 were more than the sum of the next 9 ranked countries combined (Kallbekken *et al.*, 2007). Summerton *et al.* (2015) provide a model-based macroeconomic assessment of first-mover advantage (FMA) in environmental technologies, in which a European sector becomes world leader and captures the global market for a particular technology. Defining FMA as the opportunity to capture market share by being the first to develop a new technology and establish incumbency, and using the E3MG model, Summerton *et al.* (2015) consider four scenarios where the EU is able to capture FMA in all renewables, wind and solar technologies, and motor vehicles, with the EU's FMA gradually eroding over time, and becoming zero after ten years. It is found that some of the sectors that benefit the most are also those that could be expected to lose out from higher energy prices. The model results also show that FMA could provide a small but noticeable boost to European GDP and employment, which could go some way to countering losses in production from implementing ambitious climate policy.

This is consistent with the Porter hypothesis (Porter, 1991; Porter and van der Linde, 1995),[7] which suggests that well-designed environmental regulation can spur innovation and improve competiveness by spurring innovation. Such innovation would provide a first mover in low-carbon technologies a competitive edge over their competitors. In the aforementioned *ex post* studies on the effects of the ETRs applied to the eight sectors in the four EU member states, defining that consistency with the Porter hypothesis would be if the change in the energy costs as a result of the introduction of the ETRs is lower than the savings in energy costs in a given period, Miltner and Salmons (2009) show that there is some evidence of consistency with the Porter hypothesis. Only the glass sector (26.1) shows consistency with the Porter hypothesis in all four member states, and the basic chemicals sector (24.1) in all but one. In all other sectors, two EU member states show consistency and two do not. Overall, out of 32 cases considered, 19 cases would be consistent with the Porter hypothesis, and 13 cases not. But no studies have modeled this conveyed by the early implementation of clean energy or climate policy. This also leads to the overestimate of the net carbon leakage.

[7]See Ambec *et al.* (2011) for an overview of the key theoretical and empirical insights into the Porter hypothesis to date.

5

Options to address competitiveness and leakage concerns

// pr répondre aux problèmes de compétitivité & des fuites.

Ex post analysis of the EU ETS helps us better understand competitiveness and carbon leakage impacts of the EU ETS. However, the insights from such analyses are of limited value for the future. Its relatively short time span does not allow to observe the full potential effects on the covered installations. The EU has planned more ambitious emissions reduction targets and more use of auctioning post-2012 (European Commission, 2012a; European Parliament and the Council, 2008; ICAP, 2018). Therefore, the EU ETS will become a much stricter scheme with a rising share of auctioning on the one hand and a decreasing yearly amount of the overall allowances to be handed out to industries on the other hand in the third phase 2013-2020 and beyond. Despite the European Commission's tough position in limiting the Member States' initial generous allocations for the second phase, because of the economic recession and because a number of regulatory changes regarding the cap setting and allocation of allowances kick in as of 2013, this period ended up with a huge volume of unused emissions allowances of 2.4 billion tons or more, which were be banked forward into the third phase (European Commission, 2012a).

The European Commission has recognized the negative impact of the growing surplus of allowances, proposing back-loading auction volumes of 900 million tons of allowances from 2014–2016 to 2019–2020 as a short term measure to address a growing surplus in the EU ETS (European Commission, 2012b and 2012; European Parliament and the Council, 2015). In 2015, the European Commission further established the Market Stability Reserve (MSR) as a structural measure to address the large accumulated allowance surplus. Allowances will be added to the MSR if the total number of allowances in circulation is higher than 833 million tons and re-injected to the market if the number of allowances in circulation falls below 400 million tons (European Parliament and the Council, 2015). To accelerate the reduction of the current surplus of allowances in the EU carbon market, the intake rate of the previously agreed MSR will be doubled. If the number of available emissions allowances exceeds 833 million tons, the MSR will withdraw 24% (instead of 12%) of allowances per year for the first four years starting in 2019, and surplus allowances removed from the market are placed in the MSR. As of 2023, the MSR will be capped at the number of allowances auctioned in the previous year, and excess allowances will no longer be valid (European Parliament and the Council, 2008; ICAP, 2018). In addition to these measures, the European Commission strengthens the EU ETS by means of tightening the greenhouse gas reduction target and the ETS emissions cap itself. The European Parliament and the Council (2008) in February 2018 formally approved the reform of the EU ETS for the period after 2020. The amendments to the EU ETS Directive will reduce availability of CO_2 emission allowances in the EU carbon market through increasing the annual linear reduction factor, by which the emissions cap is reduced each year from 1.74% to 2.2%, in order to comply with the EU 2030 target of reducing emissions by 43% compared to 2005 in the sectors covered by the EU-ETS. The linear reduction factor will be subject to a review in light of the goals and the stocktaking process of the Paris Agreement.

Implemented timely and effectively, these measures would help restore a higher carbon price. As would be expected, the costs of abating carbon are expected to rise in the carbon constrained EU or other countries that take comparable climate actions, relative to the rest of the non- or less carbon constrained world. While implementing strengthened ETS and other climate policies and measures is aimed to cut carbon emissions, these countries have to look for options to level the carbon costs in order to avoid putting industries exposed to a risk of carbon leakage at a competitive disadvantage vis-à-vis their trading partners.

A number of policy options have been widely discussed, and in some cases deployed to address competiveness and leakage concerns. This section discusses broad policy options to avoid putting industries exposed to a risk of carbon leakage at a competitive disadvantage,[8] while the next section focuses on border carbon adjustment measures that are incorporated in the U.S. proposed congressional climate bills to level the carbon playing field and raise great controversies.

5.1 Global approach

Clearly, the global approach is the best approach to leveling the carbon costs. The ideal solution is to have a legally binding international agreement that covers all countries, at least all major emitting economies, and sets their emissions limits. Such an agreement will enable to internalize the carbon costs and establish a global carbon price framework, thus leveling the carbon playing field. To that end, the three major milestones in international climate negotiations are the United Nations-

[8]While it is not discussed in this monograph, it should be pointed out that crediting mechanism like the clean development mechanism under the Kyoto Protocol has the potential to reduce carbon leakage and mitigate competitiveness, because it lowers the carbon price differential with non-regulating countries, which is an important driver of leakage (Kallbekken, 2007; Burniaux *et al.*, 2009). However, whether crediting mechanism reduces carbon leakage depends partly on appropriate setting of the baseline against which credits are granted (Kallbekken *et al.*, 2007). Moreover, high transaction costs of and barriers to implementation of the CDM projects, such as lack of capacity to identify and assess potential projects in host countries, risk and uncertainties associated with generating certified credits, and delays in approving CDM projects, limit its potential to lower carbon leakage, because not all potential CDM projects will be implemented. The size and nature of fossil fuel markets matter too. Trade is more regional or local for coal than for oil. CDM projects reduce fossil fuel prices in local markets. This results in increased fossil fuel use and increased carbon emissions. The CDM may then increase carbon leakage (Bollen *et al.*, 1999; Rosendahl and Strand, 2011).

sponsored climate change conferences in Kyoto in December 1997, in Copenhagen in December 2009, and the 21st session of the Conference of the Parties (COP21) to the UNFCCC in Paris in December 2015. The United Nations climate summit in Kyoto established the first legally binding climate change agreement-the Kyoto Protocol, Copenhagen aimed to succeed the Kyoto agreement, and the Paris conference reached an agreement for the post-2020 era establishing absolute, quantitative commitments for all the major economies (Zhang, 2017). For the first time, an international climate agreement known as the Paris Agreement critically specifies the long-term ambition of "holding the increase in the global average temperature to well below 2° above pre-industrial levels and to pursue efforts to limit the temperature increase to 1.5° " (UNFCCC, 2015). This is a significant strengthening compared to the earlier 2-degree target that was initially included in the Copenhagen Accord.

While the Paris Agreement establishes legally binding obligations of conduct, it does not establish a legally binding obligation for countries to actually achieve their contributions (Obergassel *et al.*, 2016). Moreover, the Paris Agreement is built on a bottom-up approach. This decentralized nature of the climate agreement means that countries have considerable flexibility in determining their own climate targets and instruments incorporated in their NDCs. Furthermore, under the Paris Agreement, countries will be mandated to submit an updated NDCs every five years, with parties expected to progress in the levels of ambition in each round in line with their national circumstances. In order to meet their existing and future targets with minimum adverse effects on their own, countries with differentiated legal and political systems are highly likely to pursue wide-ranging climate-related policies. This will well lead to large asymmetries in shadow carbon pricing across countries, and undermines the environmental effectiveness of this international climate agreement.

Establishing global sectoral agreements to supplement the Paris Agreement constitutes a robust policy to enhance carbon mitigation effect.[9] One type of such cooperative sectoral agreements aim at either

[9]See Bradley *et al.* (2007) and Meckling and Chung (2009) for further discussion

engaging specific sectors in non- or less carbon-constrained countries and eliminating, at least reducing carbon-cost differences among countries of uneven climate policies, thus restoring a level carbon playing field. Another type is to cover those sectors whose emissions are not covered by international climate agreements, such as emissions from fuels used for international aviation and maritime transport. To what extent global sectoral agreements can address competitiveness and leakage concerns depends on specific forms that such agreements take.

One form is to establish global sectoral agreements among companies within a given sector. Some industry-led agreements, e.g., the Cement Sustainable Initiative (CSI) led by the World Business Council for Sustainable Development, take this approach. The CSI gathers 24 major cement producers with operations in more than 100 countries, which collectively account for about 30% of the world's cement production. Cement production generates about 5% of global man-made CO_2 emissions, and demand for cement is forecast to continue increasing worldwide, in particular in emerging economies where much needed housing and infrastructure boosts development. Therefore, the reduction of CO_2 emissions from cement production is an urgent task for the cement sector. This CSI initiative involves capping emissions for companies operating in countries with mandatory mitigation obligations, and setting intensity-based targets for developing countries which would be differentiated to reflect national production constraints in the cement sector. The CSI member companies measure and report their CO_2 emissions following the CSI's standardized methodology documented in "*Cement CO_2 and Energy Protocol*",[10] and have been addressing CO_2 reduction since the start of the initiative by means of improving energy efficiency via modern dry-process technology, using alternative materials and fuels in the cement kiln heating process, substituting clinker with other mineral components in cement.[11] The CSI has identified

on sectoral approaches to international climate agreements.

[10]See the CSI's CO2 and Energy Accounting and Reporting Standard for the Cement Industry at http://www.wbcsdcement.org/index.php/key-issues/climate-protection/co-accounting-and-reporting-standard-for-the-cement-industry

[11]See the web site of the World Business Council for Sustainable Development's Cement Sustainable Initiative at http://www.wbcsdcement.org/.

and evaluated technologies which may contribute to increasing energy efficiency and reducing greenhouse gas emissions from global cement production by 2050, the technological potentials for improvement on energy efficiency and CO_2 emissions, boundaries and limitations, and the timeline for implementation. They could serve as the important basis for developing further technology roadmaps in the cement sector (CSI and ECRA, 2017).

The most significant and long-awaited sectoral agreements are those that aim to address emissions from international bunker fuels used for international aviation and maritime transport.[12] While emissions from international bunker fuels have been addressed under the UNFCCC process since the first meeting of the Conference of the Parties, the responsibility for reaching agreements on international transport emissions has been assigned to the International Civil Aviation Organisation (ICAO) and the International Maritime Organisation (IMO) since the Kyoto Protocol. The process to reach a global deal is slow. The EU attempted to order foreign flights into and out of the EU to buy carbon allowances under its emissions trading system in 2012. Clearly, this was an attempt to ensure a level playing field with respect to the European carriers whose emissions would be covered by the ETS. But the EU backtracked when the ICAO member countries insisted that it violated their sovereignty. So the EU agreed to exempt international flights until the start of 2017 to give ICAO time to strike a global deal. The ICAO assembly in October 2016 agreed to cap greenhouse gas emissions from global aviation from 2020 and to halve net emissions by 2050 compared to 2005 levels (ICAO, 2016a).Emissions increases after 2020 will need to be offset by implementing a global market-based measure in the form of the Carbon Offsetting and Reduction Scheme for International Aviation (CORSIA)[13] to address any annual increase in total CO_2 emissions from international civil aviation above the 2020 levels (ICAO, 2016b). Because the attempt at "carbon neutral growth" for aviation from

[12]See the UNFCCC web site on emissions from fuel used for international aviation and maritime transport at http://unfccc.int/methods/emissions_from_intl_transport/items/1057.php

[13]See the ICAO web site on the CORSIA at https://www.icao.int/environmental-protection/Pages/market-based-measures.aspx

2020 relies exclusively on offsetting, the issues remain regarding the level of participation, the transparency of the scheme, the quality of carbon offsets, environmental integrity and avoiding double counting under the CORSIA, and safeguards against the use of unsustainable aviation biofuels (ENB, 2017; Murphy, 2017). The EU has maintained limitations to the geographic scope of aviation coverage to flights within the European Economic Area until December 2023 for the ICAO's progress and clarity surrounding the content and nature of the CORSIA (ICAP, 2018). Anyhow, a global climate deal for aviation significantly increases the pressure on the IMO to act on addressing emissions from international shipping. The IMO reported that members approved a roadmap for the development of a comprehensive strategy for the reduction of greenhouse gas emissions from international shipping. The EU wants to include international shipping under its emissions trading scheme by 2023 if no ambitious and effective IMO deal is agreed by then (Abbasov, 2017; ENB, 2017). That is, the EU keeps the option of including shipping in the ETS as a leverage, and holds the IMO to account.

Sectoral agreements could also take the form of sector-specific performance standards. However, past experience shows that the negotiation of international performance standards is usually a long, if not fruitless, process requiring considerable efforts. For example, in the EU, measures to promote energy efficiency have been under serious discussion since mid-1970s, but only in the early 1990s did the EU succeed in introducing energy labelling, and the first energy efficiency standards only began to enter into force in the late 1990s (Brack *et al.*, 1999). Even if it is very difficult for the EU member countries at comparably economic and technological levels to harmonize their energy standards, the prospects for the agreements of energy efficiency or performance standards among Annex 1 countries and beyond seem remote. Indeed, developing countries led by India view strongly against sectoral performance standards. Even if common standards had eventually been established internationally, they would be expected to be driven down towards the lowest common denominator among the countries involved. Therefore, to avoid running the risk of the race to the bottom, a strong commitment by governments is required. This is even so if it is attempted to set national sectoral binding emissions targets in major economies, the stringent form of global sectoral agreements. To that end, the significantly scaled up technology transfer and deployment, financing and capacity building by industrialized countries will be crucial to encourage and enable developing countries to make strong commitments than would otherwise have been the case.

Unlike industry-led sectoral agreements, global sectoral agreements on both sectoral emissions caps and sector-specific performance standards and national sectoral binding emissions targets are subject to the UNFCCC negotiations. Provided that they are supported and enabled by finance and technology and are built upon a strong commitment by governments and the risk of a lowest common denominator can be avoided, these stringent forms of global sectoral agreements will supplement the Paris Agreement to address competitiveness and leakage concerns.

5.2 Grandfathering

Grandfathering gives out allowances freely to existing regulated entities in proportion to their historical emissions or output. In the U.S. SO_2 allowance trading program (Ellerman *et al.*, 2000) and in the first and second phases of the EU ETS, the overwhelming majority of allowances are granted for free (European Commission, 2009a).

Given that firms treat free allowances in the same way as they would do purchased allowances (e.g., Zhang and Nentjes, 1998), it is thus likely that firms pass through some, if not all, of the opportunity cost from holding allowances to consumers so that they can increase short-term profits. Pass-through rates differ significantly across sectors and among countries. Empirical studies on cost pass-through and windfall profits in the Dutch and German power sectors estimate that pass-through rates range from 60-100% for wholesale power markets (Sijm *et al.*, 2006). Alexeeva-Talebi (2011) analyses the ability of the refining sectors to pass through carbon costs to the retailers during the first phase from 2005 to 2007 by estimating a sequence of vector error correction models covering 14 EU Member States, and concludes that the full pass-through rate (100%) for opportunity costs of carbon is very likely, while the ability of producers to pass through the CO_2 opportunity cost is estimated much lower in other sectors, such as cement and iron and steel (Demailly and Quirion, 2006; Smale *et al.*, 2006). But doing so may result in a decline of market share, and thus international competitiveness concerns would not be fully alleviated (Burniaux *et al.*, 2009). Studies by Hourcade *et al.* (2007) and Smith (2008) show some evidence that this has occurred, with firms receiving free allowances trading off short-term profits and longer-term market share. In the cement sector and to a less extent in the steel sector under the EU ETS, EU producers pass through the opportunity costs of the allowances to their product prices to maintain profit margins, but face market share losses (Hourcade *et al.*, 2007).

Free allocation, as practiced in the EU ETS, can reduce incentives for relocation of production capacities in the short run, i.e., investment leakage coming from a change in the location of production capacitates. However, if free allocation is not made conditional on continued operation, companies will continue to receive free allowances. Even if they continue to operate but free allocations are not made conditional on their production level (see Section 5.3 for discussion on the updating allocation, whereby current emissions (or other activity measures) determine future allocation. This has been partly the case in the EU ETS, where emissions in the first phase affected allocation in the second phase, and production in the second phase will affect allocation in the third phase.), they can still choose to reduce production in favor of imports to keep their overall profitability. Thus unconditional free allocation does not reduce operational leakage.

Thus, free allowance allocation can compensate for carbon costs, but given the limited control over the reaction of industries unconditional free allocation will undermine its ability to address competitiveness and leakage concerns. The new closure rule under the EU ETS aims to correct this, specifying that installations that cease their operation are no longer given free allowances. Moreover, the closure rule applied since 2013 specifies that if an installation produces less than 50% of its historic activity level, its operator loses 50% of its allowances, and 75% of the allowances if it is less than 25% (European Commission, 2009b). This implies that it may be profitable to *not* close an installation.[14]

[14]The new closure rule takes the form of activity thresholds so that retrofitting an

5.3 Output-based allocation or rebates

In order to effectively address leakage, the quantity of allowances that a company receives for free needs to be tied to its production level. As the Swedish NOx charge has specified, companies above the average standards face a net liability, while those below the average get a net payment or subsidy. Since the charge was introduced in 1992, NOx emissions per unit of useful energy produced by regulated companies have declined by 50% as a result of technology adoption and innovations in physical technology and management practices. Costs of mitigation also fall significantly (OECD, 2010). The output-based allocation operates very much like tradable emissions performance standards, and thus provides similar incentives. Because additional production receives additional allowances, such an allocation is essentially a production subsidy. This will lower the marginal costs of production, and thus improve competitiveness and reduce the incentive to relocate production away from the carbon-constrained region. The Waxman-Markey bill has taken this approach, with output-based rebates for a set of energy-intensive, trade-exposed sectors, until import allowance requirements are imposed in these sectors of those countries that have not undertaken climate efforts comparable to that of the U.S. no earlier than 2020.

Output-based updating allocation has its own controversies (Quirion, 2009). This form of allocation preserves the incentive for domestic firms to reduce emissions per unit of production, but lowers the incentive for domestic firms to curtail production to meet emission targets. This is because output-based allocation acts as an implicit production subsidy, and therefore the incentives to switch from buying emission-intensive to less emission-intensive products are weakened under such an allocation. Böhringer *et al.* (2014) suggest that (1) output-based allocation continues to stimulate too much output of the emission-intensive and trade-exposed goods, whereas border carbon adjustments automatically become inactive as the coalition of regulating countries covers

inefficient plant does not reduce the amount of allowances received. This generates an incentive, for a firm operating several plants, to operate all of them at 50% at least, rather than closing the least efficient plants and operating the others close to their nominal capacity, which could be the optimal solution without these closure rules (Branger and Quirion, 2013; Climate Strategies, 2013).

the whole world, and that (2) output-based allocation triggers too much production, whereas border carbon adjustments for goods without trade exposure have little or no impacts. This would raise risk of non-compliance. Updating allocations involves periodically adjusting allocations over time to reflect changes in company's production. This can thus have the potential to provide perverse, undesirable incentives for not taking early actions or produce a negative dynamic incentive because companies anticipate future updates and adjust their decisions, e.g., by emitting more or keeping inefficient plants in operation (Neuhoff, 2008; Dröge *et al.*, 2009). Such negative effects on the company's output and innovation behaviors may drive up the cost of complying with emissions limits. Moreover, practically this approach requires an update of the allocation when production is known. The EU views output-based allocation as an *ex post* adjustment, which produces a negative dynamic incentive. Therefore, the EU ETS does not allow for output-based allocation. Rather, the EU ETS is based on production capacity when deciding free allocation, which can be seen as an intermediate one between free allocation based on historical data and output-based allocation (Quirion, 2010).

Furthermore, the effectiveness of output-based allocation in addressing carbon leakage depends on the implementation stage of the production chain. If allocation is downstream, this does not necessarily address carbon leakage. As Hourcade *et al.* (2007) show that the risk of carbon leakage is the highest for semi-finished products, producers might substitute carbon-intensive intermediate product, like clinker[15] and semi-finished steel by imports that are produced in non-carbon constrained regions. This can be avoided if allocation is upstream. However, this will eliminate incentives to use carbon pricing mechanism to reduce carbon-intensive inputs in the downstream sectors. Taking these points together, output-based allocation can be used to address carbon leakage concerns, if applied to intermediate or basic inputs (Dröge *et al.*, 2009).

[15]Clinker is the major raw material for cement and can be transported more easily than cement. But its production accounts for most of the CO_2 emitted in the manufacturing process. Therefore in the cement sector, carbon leakage is more likely to occur through clinker trade than through cement trade (Branger and Quirion, 2013).

les quotas gratuits ⇔ aux exonérations de la taxe carbone

5.4 Free allowances versus the exemptions from carbon taxes

The fears of competitiveness losses and of carbon leakage lead to proposals for sectoral exemption from the coverage of a carbon pricing. Provided that carbon pricing cannot be avoided, firms at least in some sectors would seek less detrimental options that include permitting them to pay a lower carbon tax rate than firms in other sectors in a given economy or receiving free allocation of allowances.

Free allocations of allowances would be viewed as a financial contribution conferred on the regulated firms, and would thus be regarded as a subsidy under the WTO Agreement on Subsidies and Countervailing Measures (de Cendra, 2006; James, 2009; Howse, 2010). This also applies to output-based allocation or rebates. This is an issue open to debate. There have been no agreements among analysts. Nor is any WTO jurisprudence on the issue (Stiglitz, 2006; Bhagwati and Mavroidis, 2007; Hufbauer and Kim, 2009) . However, free allocations do not seem to raise the hackles to the same extent as border adjustments to be discussed in Section 6 for at least two reasons. The first one is that countries recognize that such allocations are considered to be necessary to gain domestic acceptance of regulated firms, in particular those carbon-intensive, often well-organized industries because they are highly mobilized politically to exert greater influence on political negotiations and policy formulation. The second reason is that although grandfathering is thought of as giving implicit subsidies to some sectors, grandfathering is less trade-distorted than the exemptions from carbon pricing (Zhang, 1998b and 1999).

This point merits some explanations. Grandfathered allowances are an unconditional lump-sum transfer from the regulating government, not tied to carbon abatement efforts or production. Because they are a fixed subsidy, they have no effect on the marginal operating costs and international competitiveness of receiving firms. However, grandfathering itself implies an opportunity cost for firms receiving permits: what matters here is not how you get your permits, but what you can sell them for - that is what determines opportunity cost. Thus, even if permits are awarded gratis, firms will value them at their market price. Accordingly, the prices of products will adjust to reflect the increased scarcity of fossil fuels. This means that regardless of whether emissions permits are given out freely or are auctioned by the government, the effects on prices are expected to be the same, although the initial ownership of emissions permits differs among different allocation methods. As a result, relative prices of products will not be distorted relative to their pre-existing levels and switching of demands towards products of those firms whose permits are awarded gratis (the so-called substitution effect) will not be induced by grandfathering.

This makes grandfathering different from the exemptions from carbon taxes. In the latter case, there exist substitution effects. For example, the Commission of the European Communities (CEC) proposal for a mixed carbon and energy tax[16] provides for exemptions for the six energy-intensive industries (i.e., iron and steel, non-ferrous metals, chemicals, cement, glass, and pulp and paper) from coverage of the CEC tax on grounds of competitiveness. This not only reduces the effectiveness of the CEC tax in achieving its objective of reducing CO_2 emissions, but also makes the industries, which are exempt from paying the CEC tax, improve their competitive position in relation to those industries which are not. Therefore, there will be some switching of demand towards the products of these energy-intensive industries, which is precisely the reaction that such a tax should avoid (Zhang, 1998a and 1999). In countries such as Denmark, Norway and Sweden, where carbon/energy taxes are already implemented, energy-intensive industries are generally exempted from the taxes, either totally or partially (which leads to a large gap between effective and nominal tax rates) (Bruvoll and Larsen, 2004; Zhang and Baranzini, 2004). Since a carbon tax is intended to fall most heavily on the products of carbon-intensive industries, the exclusion of these industries from coverage of the carbon tax on the grounds of competitiveness reduces the effectiveness of the carbon tax in achieving its objective of reducing CO_2 emissions. Bruvoll

[16] As part of its comprehensive strategy to control CO_2 emissions and increase energy efficiency, a carbon/energy tax has been proposed by the CEC. The CEC proposal is that member states introduce a carbon/energy tax of US$3 per barrel oil equivalent in 1993, rising in real terms by US$1 a year to US$10 per barrel in 2000. After the year 2000 the tax rate will remain at US$10 per barrel at 1993 prices. The tax rates are allocated across fuels, with 50% based on carbon content and 50% on energy content (Zhang, 1998a).

and Larsen (2004) show that although the carbon tax implemented in Norway reached as high as US$ 51 per ton of CO_2 in 1999, exempting those sectors where the carbon tax would have been otherwise effective only led to the modest reduction in emissions of 2.3% in comparison with no carbon tax case. Moreover, while such sector exemptions from carbon taxes help maintain these sector's competitiveness, they come at a substantial cost to welfare (Hoel, 1996). As empirical studies for Germany and Canada suggest, this is because the narrowing of the tax base requires a higher tax rate for the non-exempt industries and thus increase the costs of achieving a given level of emissions reduction (Böhringer and Rutherford, 1997; Wigle, 2001). Böhringer and Rutherford (1997) find that losses associated with exemptions can be substantial, even when the share of exempted sectors in overall economic activity and carbon emissions is small.

5.5 Border carbon adjustments

As aforementioned, competitiveness and leakage concerns are also the motivations of proposals for border carbon adjustments (BCA). BCA aims to level the carbon playing field between domestic and foreign producers when they face different carbon constraints. BCA could come in the form of import taxes or the forced surrender of emissions allowances from domestic emissions trading schemes. This type of border adjustments focuses exclusively on imports, leveling the playing field for domestic consumption, but does nothing to address the competitiveness of exports in foreign markets. The border adjustment provision in the Lieberman-Warner bill works this way. BCA could also take the form of rebates for exports.[17] This type of BCA levels the playing field abroad, but still gives imports a comparative advantage at home. Full BCA combines these types of BCA, making adjustments for both imports and exports. This reflects the application of the destination principle to products, which suggests that products should be taxed in the country where they are consumed and not in the country where they are produced unless they are also consumed there. The border

[17] See Genasci (2008) for discussion on complicating issues related to how to rebate exports under a cap-and-trade regime.

adjustment provision in the Waxman-Markey bill works this way. The bill includes both rebates for few energy-intensive, trade-sensitive sectors and free emission allowances to help not to put U.S. manufacturers at a disadvantage relative to overseas competitors.

While each type of BCA has the potential to mitigate carbon leakage, the effectiveness of BCA in avoiding carbon leakage differs depending on the coverage (firm specific, sector wide or the whole economy), the source of emissions (direct emissions or both direct and indirect emissions, which refer to those emissions from both electricity generation within the country and generated from the amount of imported electricity from trading partners outside the country), the scope of adjustments (imports, exports or both) and the specific emissions intensities used for adjusting emissions embodied in the products affected. If the carbon-constrained home country is less carbon-intensive than the non-carbon constrained trading partners, taxing the carbon footprint of imports using the domestic emissions intensity is more effective at reducing carbon leakage than based on the foreign emissions intensities.

A number of modeling studies have examined the effectiveness of BCA in protecting competitiveness and avoiding carbon leakage. No general agreement has been arrived yet. On the one hand, some argue that BCA would have positive effects on environment improvements as well as competitive disadvantage offset (Majocchi and Missaglia, 2002; Veenendaal and Manders, 2008). For example, using a global partial equilibrium model of the steel industry, Mathiesen and Maestad (2004) show that border adjustments tackle leakage very effectively: a border adjustment based on average specific emissions in Non-Annex I countries can cut a carbon leakage rate to −18% from 26% without the border adjustment. Lessmann *et al.* (2009) find the influences of carbon tariffs on international cooperation significantly positive. Demailly and Quirion (2008) suggest that, based on a global partial equilibrium model of the cement industry, a carbon leakage falls from 25% without BCA to − 2% or 4%, depending on the assumed level of the BCA. Ross et al. (2009) suggest BCA an effective way for U.S. climate mitigation. Dissou and Eyland (2011) find that competitiveness would be removed by BCA in Canada. Böhringer and Rutherford (1997) suggest that BCA is the most effective leakage avoiding complement to U.S. and EU unilateral emissions reductions. Monjon and Quirion (2011) show that BCA is most effective to addressing carbon leakage under the EU ETS.

On the other hand, some studies have concluded that BCA would be ineffective either at increasing domestic competitiveness or improving global environment (Dong and Whalley, 2009a, 2009b; Elliott *et al.*, 2010). For example, Fischer and Fox (2009) suggest that BCA would not be effective at reducing global emissions. Mckibbin and Wilcoxen (2009) find modest effect of BCA to reduce leakage and to defend import-competing industries without carbon costs. Kuik and Hofkes (2010) focus on the carbon leakage avoiding effects of the EU ETS. Their results suggest that while BCA might reduce the sectoral leakage rate of the iron and steel industry, the overall leakage reduction effect is modest. The authors find an aggregate carbon leakage rate, which is 11% without BCA, decreases to 10% if BCA is based on the direct CO_2 emissions per unit of like product in the EU and to 8% if it is based on the average direct CO_2 emissions embodied per unit of foreign production of the like product. The limited impact of BCA is partly because, as aforementioned, BCA cannot prevent leakage through the fossil fuel channel, and partly the BCA limits to imports and direct emissions, and covers neither exports nor indirect emissions. Winchester *et al.* (2011) suggest that BCA results in only modest reduction in global emissions and significantly reduces welfare. Therefore they conclude that BCA is a costly means of reducing carbon leakage.

In addition, there could be potential conflicts with WTO provisions and practical difficulties associated with the implementation of BCA measures. These issues will be discussed in great detail in Section 6.

5.6 Output-based allocation or rebates versus border carbon adjustments

A number of economic modeling studies have compared which anti-leakage policy, border adjustments or output-based allocation, is more effective. The results of these studies reveal that while both policies help to protect domestic production, there is a wide disagreement on the effectiveness to limit leakage impacts (Babiker and Rutherford, 2005; Dröge *et al.*, 2009; Fischer and Fox, 2009; Quirion, 2010; Böhringer *et al.*, 2010 and 2014; Monjon and Quirion, 2011). In each case, the effectiveness depends on the objective, namely, limiting carbon leakages or mitigating production loss in the sectors affected, and differs among sectors and across countries. This last point emphasizes the importance of undertaking this kind of analysis using a highly disaggregated model.

Fischer and Fox (2009) simulate the effects of US$50 per ton of carbon emissions unilaterally implemented in several carbon-intensive sectors in the U.S. When the effectiveness is considered on avoided production loss, output-based allocation is most effective for most sectors. By contrast, full border adjustment, the sum of the import and export rebate, is most effective for iron and steel production and nonmetallic minerals when foreign embodied emissions are fully taxed at the foreign specific emissions rates. When the effectiveness refers to avoided carbon leakage, for most sectors full border adjustment is most effective at reducing global carbon emissions. However, the differences become rather small among the options considered.

Böhringer *et al.* (2010 and 2014) examine anti-leakage policies to complement the EU and U.S. unilateral policy to cut their carbon emissions by 20% relative to their baselines. In percentage changes in production and carbon leakage, their effects on the EU are bigger than on the U.S. (see Figure 5.1). One reason for this difference is that the EU is a more open economy than the U.S., having a larger share of foreign trade. This is true for both energy-intensive goods and fossil fuels. Another reason is that energy-intensive industries and power generation in the EU are less carbon-intensive than in the U.S., with both energy and carbon intensity of these sectors in the U.S. nearly doubling that in the EU (Mattoo *et al.*, 2009). Differing from the aforementioned findings of Fischer and Fox (2009), Böhringer *et al.* (2010 and 2014) find that full border adjustment is most effective at both mitigating production loss (see Figure 5.2) and reducing carbon leakages (see Figure 5.1), in particular for the EU where such effects are much bigger for the EU as a whole and all energy-intensive sectors. Moreover, import tariffs are much more effective than export rebates in protecting competitiveness and avoiding carbon leakage (see Figure 5.1). Among the three anti-leakage policies, namely, import tariffs, full border adjustment and output-based allocation, output-based allocation is not only least effective at deterring carbon leakage, but also most costly, although it still provides some cost savings over unilateral climate policy when the size of the abatement coalition is confined to industrialized countries. As the coalition expands,

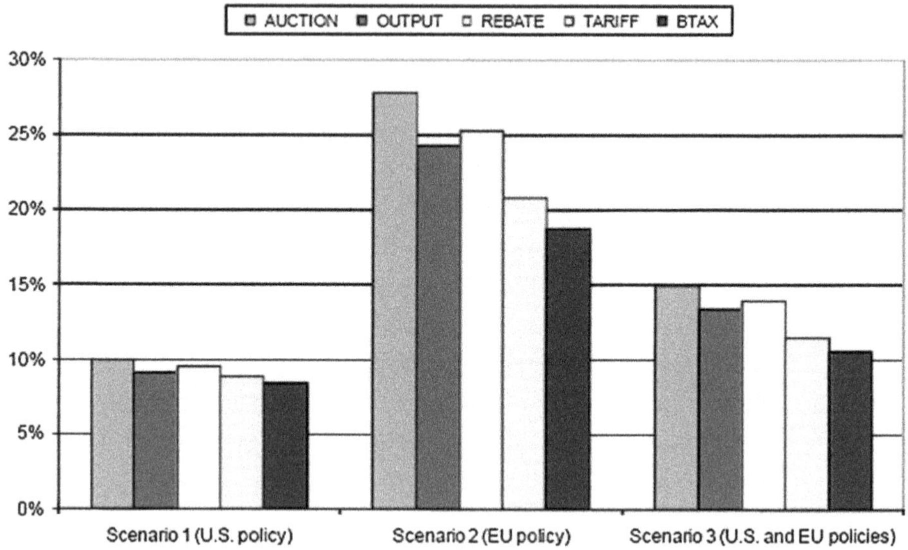

Figure 5.1: Global leakage effects of U.S. and/or EU CO2 emissions reductions
Source: Böhringer *et al.* (2010).

however, output-based allocation induces excess costs as it maintains distortionary subsidies to energy-intensive, trade-exposed production, whereas the cost savings through leakage reduction declines (Böhringer *et al.*, 2014).

Monjon and Quirion (2011) compare a variety of variants of border adjustments and output-based allocations to cap emissions under the EU ETS in 2016 at 85% of their 2005 levels. They find that output-based allocation is more effective than border adjustment to mitigate production losses of the sectors affected by the EU ETS. The same conclusion is reached by Rivers (2010), which examines a 20% reduction in CO_2 emissions from 2006 levels by 2020 in Canada. While this finding generally shares that from Fischer and Fox (2009), the results of the two studies differ when it comes to specific sectors. For example, in the iron and steel sector, Fischer and Fox (2009) show that taxing foreign embodied emissions at the foreign specific emissions rates helps to avoid U.S. production loss more effectively than allocation-based home rebates. This is just opposite to that of Monjon and Quirion (2011) for the EU. The specific emissions factors used are attributable

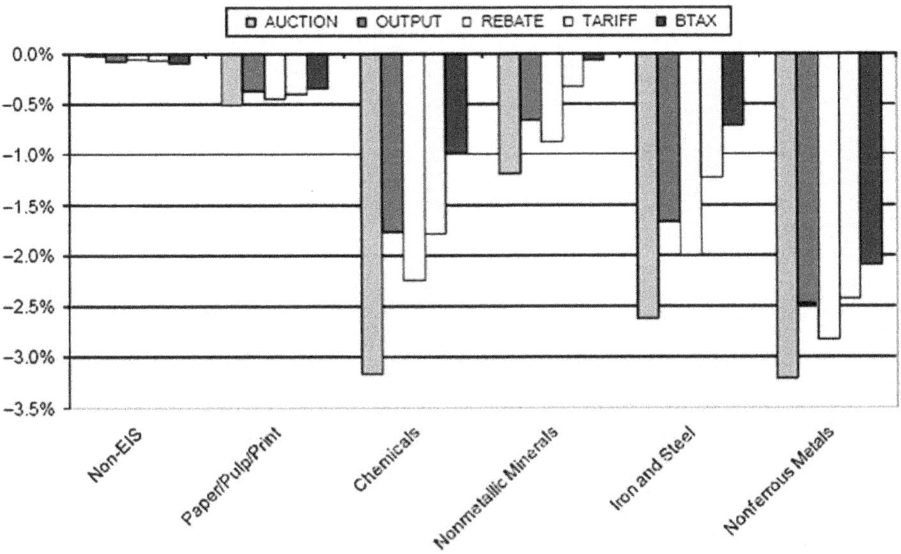

Figure 5.2: Effects of EU climate policies on EU production
Source: Böhringer *et al.* (2010).

to this difference. In the EU study, foreign emission intensity relative to the EU is 137%, whereas foreign emission intensity relative to the U.S. is 295%.

When it comes to tackling carbon leakage, as shown in Figure 5.3, Monjon and Quirion (2011) suggest that border adjustment is more effective than output-based allocation. While this is generally in line with the U.S. study, the differences in the effectiveness among the options considered are big, in comparison with the U.S. study which is impossible to rank the options. This is mainly because border carbon adjustment reduces the overall consumption of carbon-intensive steel, aluminum and cement in the EU, and therefore cuts their imports from outside the EU as well as their production and CO_2 emissions. Moreover, in most sectors, the fact that EU installations have lower specific emissions than those from outside the EU help them to win some market shares on European markets, because they face a lower increase in production cost than their foreign competitors. This will help to further reduce imports and hence emissions from outside the EU. The two variants of particularly interesting are ones that closely resemble

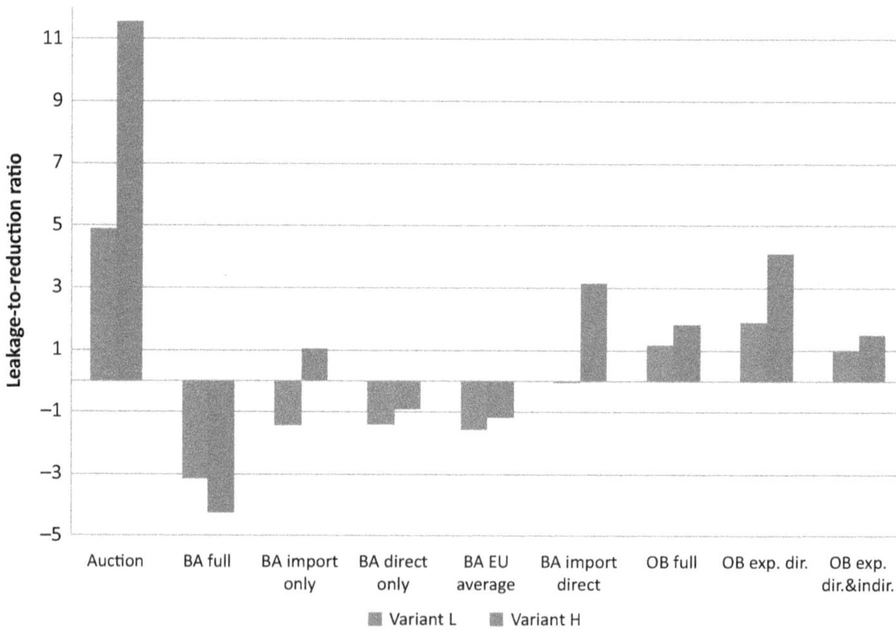

Figure 5.3: Aggregate carbon leakage rate
Source: Monjon and Quirion (2011).

the climate debates in the EU and the U.S.. In the third phase of the EU ETS, all allowances will be auctioned for electricity generation from 2013 onwards (with a transitional period for the ten new member states that are given the option of exempting themselves from the full auctioning rule and continuing to allocate a limited number of emission allowances to power plants for free until 2019) but industries deemed exposed to a significant risk of carbon leakage will continue to receive free allowances (European Commission, 2009a, 2009b and 2011a). The scenario labeled as *OB exposed direct*, in which auctioning of allowances is for electricity generation and output-based allocation is for direct emissions in exposed sectors, simulates this case. Moreover, in the EU and the U.S., the most discussed options for a border carbon adjustment focus on imports and direct emissions. The variant labeled as *BA import direct* simulates this case. The aforementioned conclusion of the superiority of BA over output-based allocation for tackling leakage holds.

In a recent study, Böhringer *et al.* (2017b) show that combining output-based rebating with a consumption tax for emission-intensive and tradeexposed (EITE) goods can be equivalent with border carbon adjustments, provided that the consumption tax for an EITE good is equal to the output-based rebating rate. However, even if previous studies suggest that output-based allocation generally leads to a higher carbon leakage than border adjustment, leakage remains very limited. As shown in Figure 5.3, under the most efficient variant with auctioning in the power sector and output-based allocation in the cement, steel and aluminum sectors for both direct and indirect emissions, the aggregate carbon leakage rate falls to around 1–2% based on high and low values of Armington elasticities. This suggests that if leakage is considered a serious issue and that border adjustments should not be applied for some reason related to international climate change negotiations or to WTO provisions, auctioning in the power sector and output-based allocation in manufacturing industry for both direct and indirect emissions, can serve as a second-best option (Monjon and Quirion, 2011). Supplementing output-based rebating with a consumption tax also constitutes a robust policy to mitigate carbon leakage. Compared with output-based rebating stand-alone, it improves costeffectiveness of unilateral climate policy; compared with border carbon adjustments, it limits the risks of detrimental trade disputes (Böhringer *et al.*, 2017b).

5.7 Carbon leakage in the context of China's pilot carbon trading schemes

Up to now, we provide a comprehensive review of carbon leakage and the effectiveness of anti-leakage policies associated with differentiated climate abatement commitments among countries. The review clearly shows that differentiated climate abatement commitments among countries have raised great concern about carbon leakage. China has launched the seven pilot carbon trading schemes in the capital Beijing, the business hub of Shanghai, the sprawling industrial municipalities of Tianjin and Chongqing, the manufacturing center of Guangdong province on the southeast coast, Hubei province, home of Wuhan Iron and Steel, Shenzhen, the Chinese Special Economic Zone and across the border from Hong Kong (Zhang, 2015a and 2015b; ICAP, 2018). Can the findings derived from the international context be directly transplanted into the domestic context of China? At first glance, carbon leakage in the domestic context is of great concern just like carbon leakage concern in the international context. But on second thought, it may be not that severe as it is initially thought. This is mainly related to the characteristics of either specific trading design of these pilot schemes or the Chinese economy as discussed below and in Zhang (2015a).

First, the sectors covered are hard to relocate. For example, the first trading phase in Guangdong only covers the four major emitting sectors of power, iron and steel, petrochemical, and cement (GPDRC, 2013). The enterprises in these sectors are capital-intensive, and need approval of the National Development and Reform Commission (NDRC) or are subject to national restrictions on the overall capacity given that their production already exceeds demand. Thus, they are not that ease to relocate from one region to another. This is very much like the issue of "footlooseness" in the U.S. industries: those industries with the largest pollution abatement costs also happen to be the least geographically mobile, or "footloose" and the competitiveness effect of environmental regulations on the least footloose industries is small (Ederington *et al.*, 2005).

Second, all pilots allocate all or the majority of allowances for free, and allocations for some sectors covered are based on emissions intensities. The free allocations of allowances on the basis of emissions intensities allow emissions caps to be *ex post* adjusted with real output, thus mitigating negative economic effects. In addition, some *ex post* adjustment mechanism is built to mitigate potentially significant increase in compliance cost. For example, the Hubei pilot specifies that if the yearly verified emissions of one covered entity exceed its cap by 20%, or 200000 tCO_2, then the extra emissions will be covered by the government allowance reserve, which is capped at 10% of total amount of allowances (HPG, 2014).

Third, differing from the EU ETS, indirect emissions from electricity generation are covered in all of the seven pilot schemes, and this design feature could help to reduce carbon leakage in two ways. The first way is to cut carbon leakage from the increased electricity imports if no indirect emissions are covered. For a region like Beijing, over 60% of electricity consumption is imported from other regions. If indirect emissions associated with imported electricity are not covered, then a significant amount of emissions in this region are not covered. The region would import more electricity instead of producing electricity on its own, thus leading to more carbon emissions in other regions than what would otherwise be the case. Covering indirect emissions caused from the amount of imported electricity would reduce the potential of carbon leakage. The second way is to cut downstream companies' potential shift to electricity consumption. As discussed in Zhang (2014), electricity tariffs in China have remained controlled by the central government and still remain flat and regulated. As such, the increased carbon costs of power generators to comply with the carbon or energy limits cannot be passed through to the downstream energy consumers. If indirect emissions are not covered, then downstream companies could reduce their emissions by means of replacing fossil fuel consumption with electricity consumption, provided that they found this shift economically profitable.

Fourth, the prevailing findings from the pollution haven literature are that environmental regulations have a small to negligible impact on relocations (Ederington *et al.*, 2005; Oikonomou *et al.*, 2006), not to mention that carbon cost is only a small portion of the overall costs, at least in the short term. Putting a price on carbon through carbon trading affects a firm's competitiveness by changing its relative production costs. However, relative production costs are determined not only by pollution regulation alone and pollution regulation may not be an important determinant of costs (Copeland and Taylor, 2003). The evidence that why environmental-led relocations were not observed in the international context indicates that differences in regional pollution policy are only one of the many factors that affect the firm's competitiveness and relocation (Jaffe *et al.*, 1995; Levinson and Taylor, 2008). The cost associated with this pollution abatement factor represents a small fraction of costs compared to other costs or barriers which still favor production in industrialized countries (Oikonomou *et al.*, 2006; Ederington *et al.*, 2005): tariffs, transport costs, labor productivity, the extent of agglomeration economies of an industry, the fixed plant costs, volatility in exchange rate, political risk, etc. If other factors outweigh the effects of carbon abatement policy on comparative advantage, then relocation would not occur to regions with weak carbon regulation.

Fifth, each province or region in China has mandatory emissions targets, and these targets are considered, at least perceived, comparable among comparable regions. In taking on emissions commitments, each province or equivalent has several rounds of bargaining with the NDRC. It would be hard for one province to make a deal with the NDRC when it finds that another province in a comparable position does not make a comparable commitment. Indeed, the bargaining process will ensure that each province takes on stringent energy-saving and pollution-cutting commitments up to its expectation. This is very different in the international context, where commitments across countries, in particular commitments by developed countries and developing countries, are widely perceived as incomparable, although such a difference is not necessarily irrational.

Sixth,in any case, it is not easy to relocate enterprises in a given region to other regions in China. There are a lot of restrictions, regardless of carbon trading. The aforementioned first point illustrates that by carbon trading design itself it is not easy to relocate enterprises in a covered region to other regions. Even without consideration of carbon trading, relocation is not easy either. Just like elsewhere in the world, local governments in China are responsible for not only providing public services, but also for promoting their local economies. However, given that the central government only accounted for less than 25% of the country's total government expenditure but received over 50% of the total government revenue in China since China adopted the tax-sharing system in 1994, local governments increasingly face the burden of rapidly growing expenditures for culture and education, supporting agricultural production, social security subsidiary, urban infrastructure and urban development, etc. (Zhang, 2008 and 2010). To enable to pay their expenditure, local governments have little choice but to focus on local development and GDP. That will in turn enable them to enlarge their tax revenue by collecting urban maintenance and development tax, contract tax, arable land occupation tax, urban land use tax, etc. Partly because of the drawback of the existing tax system in China, which separates the origin of tax from the collecting place,[18] local governments even intervene mergers, acquisitions and relocations of enterprises in order to prevent losses of revenues and negative effect of economic growth, even if these enterprises often are small-sized, less efficient and highly polluting, and the regions would benefit from mergers, acquisitions and relocations in terms of improved environmental quality and sustained economic development.

[18]Based on the existing tax system in China, the origin of tax is different from the collecting place. Enterprises are mandated to pay the business taxes to the places where they are registered, not to the places where these taxes occur. Take the project of natural gas transmission from West to East China as a case. When natural gas is transmitted from the western part of China to Shanghai, 40% of the business tax revenues are in Shanghai where the enterprises are registered, and 60% of the revenues go to the central government. But more than ten provinces along this transmission line hardly receive any business revenues, although they spend a lot of money for services. With some provisional measures on the distribution of business revenues among sub-entities across provincial borders implemented since 2008, there has been some progress but this issue has not been completely settled (Wu, 2014).

6

The WTO consistency, the effectiveness and methodological challenges of border carbon adjustment measures

In the Paris Agreement context, to date, Mexico is the only country to explicitly refer to possibility of border carbon adjustments in its NDC. The Mexican NDC states that its 25% emissions reduction commitment could be increased to 40%, subject to a number of conditions, including border carbon adjustments. While only one NDC includes an explicit reference to BCAs, this does not mean that others may not consider it (Brandi, 2017).

Indeed, border adjustment measures in the form of emissions allowance requirements (EAR) as the most concrete unilateral trade measure were incorporated in the U.S. proposed cap-and-trade regime to level the carbon playing field. If improperly implemented, such measures could disturb the world trade order and trigger a trade war. Because of these potentially far-reaching impacts, this section focuses on this type of unilateral border adjustment. It required importers to acquire and surrender emissions allowances corresponding to the embedded carbon contents in their goods from countries that had not taken climate actions comparable to that of the importing country. My discussion is mainly on the legality of unilateral EAR under the WTO rules.

6.1 Proposed border adjustment measures in the U.S. climate legislations

The notion of border carbon adjustments is not an American invention. The idea of using BCA to address the competitiveness concerns as a result of differing climate policy was first floated in the EU, in response to the U.S. withdrawal from the Kyoto Protocol. Dominique de Villepin, the then French prime minister, proposed in November 2006 for carbon tariffs on goods from countries that had not ratified the Kyoto Protocol. He clearly had the U.S. in mind when contemplating such proposals aimed to bring the U.S. back to the table for climate negotiations. However, Peter Mandelson, the then EU trade commissioner, dismissed the French proposal as not only a probable breach of trade rules but also "not good politics" (Bounds, 2006). As a balanced reflection of the divergent views on this issue, the European Commission had suggested that it could implement a "carbon equalization system ... with a view to putting installations from the Community which are at a significant risk of carbon leakage and those from third countries on a comparable footing. Such a system could apply requirements to importers that would be no less favourable than those applicable to installations within the European Union, for example by requiring the surrender of allowances" (European Commission, 2008, p. 8). In light of this, various proposals about carbon equalization systems at the border have been put forward, the most noticeable linked to French president Nicolas Sarkozy's proposal for "a carbon tax at the borders of Europe." President Sarkozy renewed such a call for a European carbon tax on imports when unveiling the details of France's controversial national carbon tax of € 17 per ton of CO_2 emissions. He defended his position by citing comments from the WTO that such a tax could be compatible with its rules and referring to a similar border carbon adjustment provision under the Waxman-Markey bill in the U.S. House to be discussed in the next two sections, arguing that "I don't see why the US can do it and Europe cannot (Hollinger, 2009, p. 5) and that not to do so would amount to "massive aid to relocations" (*The Economist*, 2009, p. 80). So far, while the EU has considered the possibility of imposing a border allowance adjustment should serious leakage issues arise in the future, it has put this option on hold at least until 2012. The European Commission has proposed using temporary free allocations to address competitiveness concerns in the interim. Its aim is to facilitate a post-2020 climate negotiation while keeping that option in the background as a last resort.

Interestingly, the U.S. legislators had not only embraced such BCA measures that they opposed in the past, but had also focused on their design issues in more details. In the U.S. Senate, the Boxer Substitute of the Lieberman-Warner Climate Security Act (S. 3036) mandated that starting from 2014 importers of products covered by the cap-and-trade scheme would have to purchase emissions allowances from an International Reserve Allowance Program if no comparable climate action were taken in the exporting country. Least developed countries and countries that emit less than 0.5% of global greenhouse gas emissions (i.e., those not considered significant emitters) would be excluded from the scheme. Given that most carbon-intensive industries in the U.S. run a substantial trade deficit (Houser *et al.*, 2008), this proposed EAR clearly aimed to level the carbon playing field for domestic producers and importers. In the U.S. House of Representatives, the American Clean Energy and Security Act of 2009 (H.R. 2998), sponsored by Reps. Henry Waxman (D-CA) and Edward Markey (D-MA), was narrowly passed on 26 June 2009, which was not passed in the U.S. Senate and was thus not adopted. The so-called Waxman-Markey bill sets up an "International Reserve Allowance Program" whereby U.S. importers of primary emission-intensive products from countries having not taken "greenhouse gas compliance obligations commensurate with those that would apply in the United States" would be required to acquire and surrender carbon emissions allowances. The EU by any definition would pass this comparability test, because it has taken under the Kyoto Protocol and is going to take in its follow-up regime much more ambitious climate targets than U.S. Because all other remaining Annex 1 countries but the U.S. had accepted mandatory emissions targets under the Kyoto Protocol, these countries would likely pass the comparability test as well, which exempts them from EAR under the U.S. cap-and-trade regime. While France targeted the American goods, the U.S. EAR clearly targeted major emerging economies, such as China and India.

examen par l'OMC des projets lois s/ climat du Congrès

6.2 WTO scrutiny of U.S. congressional climate bills

The import emissions allowance requirement was a key part of the Lieberman-Warner Climate Security Act of 2008, and would re-appear again as the U.S. lawmakers debate and vote any climate change bill in the future. Moreover, concerns raised in the Lieberman-Warner bill seemed to have provided references to writing relevant provisions in the Waxman-Markey bill to deal with the competitiveness concerns. For these reasons, I start with the Lieberman-Warner bill.

A proposal first introduced by the International Brotherhood of Electrical Workers (IBEW) and American Electric Power (AEP) in early 2007 would require importers to acquire emission allowances to cover the carbon content of certain products from countries that do not take climate actions comparable to that of the U.S. (Morris and Hill, 2007). The original version of the Lieberman-Warner bill incorporated this mechanism, threatening to punish energy-intensive imports from developing countries by requiring importers to obtain emission allowance, but only if they had not taken comparable actions by 2020, eight years after the effective start date of a U.S. cap-and-trade regime begins. It was argued that the inclusion of trade provisions would give the U.S. additional diplomatic leverage to negotiate multilaterally and bilaterally with other countries on comparable climate actions. Should such negotiations not succeed, trade provisions would provide a means of leveling the carbon playing field between American energy-intensive manufacturers and their competitors in countries not taking comparable climate actions. Not only would the bill have imposed an import allowance purchase requirement too quickly, it would have also dramatically expanded the scope of punishment: almost any manufactured product would potentially have qualified. If strictly implemented, such a provision would pose an insurmountable hurdle for developing countries (*The Economist*, 2008a).

It should be emphasized that the aim of including trade provisions is to facilitate negotiations while keeping open the possibility of invoking trade measures as a last resort. The latest version of the Lieberman-Warner bill had brought the deadline forward to 2014 to gain business and union backing.[19] The inclusion of trade provisions might be considered the "price" of passage for any U.S. legislation capping its greenhouse gas emissions. Put another way, it is likely that no climate legislation can move through U.S. Congress without including some sort of trade provisions. An important issue on the table was the length of the grace period to be granted to developing countries. While, from the legal, political economy and international climate negotiation perspectives, many factors need to be taken into consideration (Haverkamp, 2008), further bringing forward the imposition of allowance requirements to imports was rather unrealistic, given the already very short grace period ending 2019 in the original version of the bill. It should be noted that the Montreal Protocol on Substances that Deplete the Ozone Layer grants developing countries a grace period of 10 years (Zhang, 2000). Given that the scope of economic activities affected by a climate regime is several orders of magnitude larger than those covered by the Montreal Protocol, if legislation incorporates border adjustment measures (put the issue of their WTO consistency aside), in my view, they should not be invoked for at least 10 years after mandatory U.S. emission targets take effect.

Moreover, unrealistically shortening the grace period granted before resorting to the trade provisions would increase the uncertainty of whether the measure would withstand a challenge by U.S. trading partners before the WTO. As the ruling in the Shrimp-Turtle dispute indicates (see Box 3), for a trade measure to be considered WTO-consistent, a period of good-faith efforts to reach agreements among the countries concerned is needed before imposing such trade measures. Put another way, trade provisions should be preceded by major efforts to negotiate with partners within a reasonable timeframe. Furthermore, developing countries need a reasonable length of time to develop and operate national climate policies and measures. Take the establishment of an emissions trading scheme as a case in point. Even for the U.S. SO_2 Allowance Trading Program, the entire process from the U.S. Environmental Protection Agency beginning to compile the data for its

[19]This is in line with the IBEW/AEP proposal, which requires U.S. importers to submit allowances to cover the emissions produced during the manufacturing of those goods two years after U.S. starts its cap-and-trade program (McBroom, 2008).

allocation database in 1989 to publishing its final allowance allocations in March 1993 took almost four years. For the first phase of the EU Emissions Trading Scheme, the entire process took almost two years from the EU publishing the Directive establishing a scheme for greenhouse gas emission allowance trading on 23 July 2003 to it approving the last national allocation plan for Greece on 20 June 2005. For developing countries with very weak environmental institutions and that do not have dependable data on emissions, fuel uses and outputs for installations, this allocation process is expected to take much longer than what experienced in the U.S. and the EU (Zhang, 2007).

Box 2 Core General Agreement on Tariffs and Trade (GATT) principles

GATT Article 1 ('most favored nation' treatment): WTO members not allowed to discriminate against like imported products from other WTO members

GATT Article III ('national treatment'): Domestic and like imported products treated identically, including any internal taxes and regulations

GATT Article XI ('elimination of quantitative restrictions'): Forbids any restrictions (on other WTO members) in the form of bans, quotas or licenses

GATT Article XX

"Subject to the requirement that such measures are not applied in a manner which would constitute a means of arbitrary or unjustifiable discrimination between countries where the same conditions prevail, or a disguised restriction on international trade, nothing in this

Agreement shall be constructed to prevent the adoption or enforcement by any contracting party of measures...

(b) **necessary** to protect human, animal or plant life or health;

...

(g) **relating to** the conservation of exhaustible natural resources if such measures are made effective in conjunction with restrictions on domestic production or consumption; ..."

The **threshold for (b) is higher than for (g)**, because, in order to fall under (b), the measure must be "necessary", rather than merely "relating to" under (g).

Box 3 Implications of the findings of the WTO Shrimp-Turtle dispute

To address the decline of sea turtles around the world, in 1989 the U.S. Congress enacted Section 609 of Public Law 101-162 to authorize embargoes on shrimp harvested with commercial fishing technology harmful to sea turtles. The U.S. was challenged in the WTO by India, Malaysia, Pakistan and Thailand in October 1996, after embargoes were leveled against them. The four governments challenged this measure, asserting that the U.S. could not apply its laws to foreign process and production methods. A WTO Dispute Settlement Panel was established in April 1997 to hear the case. The Panel found that the U.S. failed to approach the complainant nations in serious multilateral negotiations before enforcing the U.S. law against those nations. The Panel held that the U.S. shrimp embargo was a class of measures of processes-and-production-methods type and had a serious threat to the multilateral trading system because it conditioned market access on the conservation policies of foreign countries. Thus, it cannot be justified under GATT Article XX. However, the WTO Appellate Body overruled the Panel's reasoning. The Appellate Body held that a WTO member requires from exporting countries compliance, or adoption of, certain policies prescribed by the importing country does not render the measure inconsistent with the WTO obligation. Although the Appellate Body still found that the U.S. shrimp embargo was not justified under GATT Article XX, the decision was not on the grounds that the U.S. sea turtle law itself was inconsistent with GATT. Rather, the ruling was on the grounds that the application of the law constituted "arbitrary and unjustifiable discrimination" between WTO members (WTO, 1998). The WTO Appellate Body pointed to a 1996 regional agreement reached at the U.S. initiation, namely the Inter-American Convention on Protection and Conservation of Sea Turtles, as evidence of the feasibility of such an approach (WTO, 1998; Berger, 1999). Here, the Appellate Body again advanced the standing of multilateral environmental treaties (Zhang, 2004; Zhang and Assunção, 2004). Thus, it follows that this trade dispute under the WTO may have been interpreted as a clear preference for actions taken pursuant to multilateral agreements and/or negotiated through international cooperative arrangements, such as the Kyoto Protocol and its successor. However, this interpretation should be viewed with great caution, because there is no doctrine of *stare decisis* (namely, "stand by things already established by prior decisions") in the WTO; the GATT/WTO panels are not bound by previous panel decisions (Zhang and Assunção, 2004). Moreover, the WTO Shrimp-Turtle dispute settlement has a bearing on the ongoing discussion on the "comparability" of climate actions in a post-2012 climate change regime. The Appellate Body found that when the U.S. shifted its standard from requiring measures essentially the same as the U.S. measures to "the adoption of a program *comparable* in effectiveness", this new standard would comply with the WTO disciplines (WTO, 2001, paragraph 144). Some may view that this case opens the door for U.S. climate legislation that bases trade measures on an evaluation of the comparability of climate actions taken by other trading countries. Comparable action can be interpreted as meaning action comparable in effect as the "*comparable* in effectiveness" in the Shrimp-Turtle dispute. It can also be interpreted as meaning "the comparability of efforts". The Bali Action Plan adopts the latter interpretation, using the terms comparable as a means of ensuring that developed countries undertake commitments comparable to each other (Zhang, 2009).

In the case of a WTO dispute, the question will arise whether there are any alternatives to trade provisions that could be reasonably expected to fulfill the same function but are not inconsistent or less inconsistent with the relevant WTO provisions. Take the GATT Thai cigarette dispute as a case in point. Under Section 27 of the Tobacco Act of 1966, Thailand restricted imports of cigarettes and imposed a higher tax rate on imported cigarettes when they were allowed on the three occasions since 1966, namely in 1968-70, 1976 and 1980. After consultations with Thailand failed to lead to a solution, the U.S. requested in 1990 the Dispute Settlement Panel to rule on the Thai action on the grounds that it was inconsistent with Article XI:1 of the General Agreement; was not justified by the exception under Article XI:2(c), because cigarettes were not an agricultural or fisheries product in the meaning of Article XI:1; and was not justified under Article XX(b) because the restrictions were not necessary to protect human health, i.e. controlling the consumption of cigarettes did not require an import ban. The Dispute Settlement Panel ruled against Thailand. The Panel found that Thailand had acted inconsistently with Article XI:1 for having not granted import licenses over a long period of time. Recognizing that XI:2(c) allows exceptions for fisheries and agricultural products if the restrictions are necessary to enable governments to protect farmers and fishermen who, because of the perishability of their produce, often could not withhold excess supplies of the fresh product from the market, the Panel found that cigarettes were not "like" the fresh product as leaf tobacco and thus were not among the products eligible for import restrictions under Article XI:2(c). Moreover, the Panel acknowledged that Article XX(b) allowed contracting parties to give priority to human health over trade liberalization. The Panel held the view that the import restrictions imposed by Thailand could be considered to be "necessary" in terms of Article XX(b) only if there were no alternative measure consistent with the General Agreement, or less inconsistent with it, which Thailand could reasonably be expected to employ to achieve its health policy objectives. However, the Panel found that the Thai import restriction measure was not necessary because Thailand could reasonably be expected to take strict, non-discriminatory labelling and ingredient disclosure regulations and to ban all the direct and indirect advertising, promotion and sponsorship of cigarettes to ensure the quality and reduce the quantity of cigarettes sold in Thailand. These alternative measures are considered WTO-consistent to achieve the same health policy objectives as Thailand now pursues through an import ban on all cigarettes whatever their ingredients (GATT, 1990). Simply put, in the GATT Thai cigarette dispute, the Dispute Settlement Panel concluded that Thailand had legitimate concerns with health but it had measures available to it other than a trade ban that would be consistent with the General Agreement on Tariffs and Trade (e.g. bans on advertising) (GATT, 1990).

Indeed, there are alternatives to resorting to trade provisions to protect the U.S. trade-sensitive and energy-intensive industries during a period when the U.S. was taking good-faith efforts to negotiate with trading partners on comparable actions. One way to address competitiveness concerns is to initially allocate free emission allowances to those sectors vulnerable to global competition, either totally or partially.[20] Bovenberg and Goulder (2002) found that giving out about 13% of the allowances to fossil fuel suppliers freely instead of auctioning in an emissions trading scheme in the U.S. would be sufficient to prevent their profits from falling under a carbon-constrained policy, relative to the case of no emissions constraints.

There is no disagreement that the allocation of permits to emissions sources is a politically contentious issue. Grandfathering, or at least partially grandfathering, helps these well-organized, politically highly-mobilized industries or sectors to save considerable expenditures and thus increases the political acceptability of an emissions trading scheme, although it leads to a higher economic cost than a policy where the allowances are fully auctioned.[21] This explains why the sponsors

[20]To be consistent with the WTO provisions, foreign producers could arguably demand the same proportion of free allowances as U.S. domestic producers in case they are subject to border carbon adjustments.

[21]In a second-best setting with pre-existing distortionary taxes, if allowances are auctioned, the revenues generated can then be used to reduce pre-existing distortionary taxes, thus generating overall efficiency gains. Parry *et al.* (1999), for example, show that the costs of reducing U.S. carbon emissions by 10% in a second-best setting with pre-existing labor taxes are five times more costly under a grandfathered carbon permits case than under an auctioned case. This is because the

of the American Clean Energy and Security Act of 2009 had to make a compromise amending the Act to auction only 15% of the emission permits instead of the initial proposal for auctioning all the emission permits in a proposed cap-and-trade regime. This change allowed the Act to pass the U.S. House of Representatives Energy and Commerce Committee in May 2009. Moreover, as discussed in Section 5.2, although grandfathering is thought of as giving implicit subsidies to these sectors, grandfathering is less trade-distorted than the exemptions from carbon taxes (Zhang, 1998b and 1999), which means that partially grandfathering is even less trade-distorted than the exemptions from carbon taxes.

The import allowance requirement approach would distinguish between two otherwise physically identical products on the basis of climate actions in place in the country of origin. This discrimination of like products among trading nations would constitute a prima facie violation of WTO rules. To pass WTO scrutiny of trade provisions, the U.S. is likely to make reference to the health and environmental exceptions provided under GATT Article XX (see Box 2). This Article itself is the exception that authorizes governments to employ otherwise GATT-illegal measures when such measures are necessary to deal with certain enumerated public policy problems. The GATT panel in Tuna/Dolphin II concluded that Article XX does not preclude governments from pursuing environmental concerns outside their national territory, but such extra-jurisdictional application of domestic laws would be permitted only if aimed *primarily* (emphasis added) at having a conservation or protection effect (GATT, 1994; Zhang, 1998b). The capacity of the planet's atmosphere to absorb greenhouse gas emissions without adverse impacts is an 'exhaustible natural resource.' Thus, if countries take measures on their own including extra-jurisdictional application *primarily* to prevent the depletion of this 'exhaustible natural resource,'

policy where the permits are auctioned raises revenues for the government that can be used to reduce pre-existing distortionary taxes. By contrast, in the former case, no revenue-recycling effect occurs, since no revenues are raised for the government. However, the policy produces the same tax-interaction effect as under the latter case, which tends to reduce employment and investment and thus exacerbates the distortionary effects of pre-existing taxes (Zhang, 1999).

such measures will have a good justification under GATT Article XX. Along this reasoning, if the main objective of trade provisions is to protect the environment by requiring other countries to take actions comparable to that of the U.S., then mandating importers to purchase allowances from the designated special international reserve allowance pool to cover the carbon emissions associated with the manufacture of that product is debatable. To increase the prospects for a successful WTO defence, I think that trade provisions can refer to the designated special international reserve allowance pool, but may not do so without adding "or equivalent." This allows importers to submit equivalent emission reduction units that are not necessarily allowances but are recognized by international treaties to cover the carbon contents of imported products.

Clearly, these concerns raised in the Lieberman-Warner bill had shaped relevant provisions in the Waxman-Markey bill to deal with the competitiveness and leakage concerns. Accordingly, the Waxman-Markey bill had avoided all the aforementioned controversies raised in the Lieberman-Warner bill. Unlike the EAR in the Lieberman-Warner bill which focused exclusively on imports into the U.S., but did nothing to address the competitiveness of U.S. exports in foreign markets, the Waxman-Markey bill included both rebates for few energy-intensive, trade-sensitive sectors[22] and free emission allowances to help not to put U.S. manufacturers at a disadvantage relative to overseas competitors. Unlike the Lieberman-Warner bill in the U.S. Senate, the Waxman-Markey bill also gave China, India and other major developing nations time to enact their climate-friendly measures. Under the Waxman-Markey bill, the International Reserve Allowance Program may not begin before 1 January 2025. The U.S. president may only implement an International Reserve Allowance Program for sectors producing primary products. While the bill called for a "carbon tariff" on imports, it very much framed that measures as a last resort that a U.S. president could impose at his or her discretion regarding border adjustments or tariffs. However, in the middle of the night before the vote on 26 June 2009, a provision was inserted in this House bill that required the President,

[22]See Genasci (2008) for discussion on complicating issues related to how to rebate exports under a cap-and-trade regime.

starting in 2020, to impose a border adjustment - or tariffs - on certain goods from countries that did not act to limit their greenhouse gas emissions. The president can waive the tariffs only if he received explicit permission from the U.S. Congress (Broder, 2009). The last-minute changes in the bill changed a Presidential long-term back-up option to a requirement that the President put such tariffs in place under the specified conditions. Such changes significantly changed the spirit of the bill, moving it considerably closer to risky protectionism. While praising the passage of the House bill as an "extraordinary first step," president Obama opposed a trade provision in that bill.[23] The carbon tariff proposals had also drawn fierce criticism from China and India. Without specific reference to the U.S. or the Waxman-Markey bill, China's Ministry of Commerce said in a statement posted on its website that proposals to impose "carbon tariffs" on imported products violated the rules of the World Trade Organization. That would enable developed countries to "resort to trade in the name of protecting the environment." The carbon tariff proposal run against the principle of "common but differentiated responsibilities," the spirit of both the UNFCCC and the Kyoto Protocol.[24] This would neither help strengthen confidence that the international community can cooperate to handle the (economic) crisis, nor help any country's endeavors during the climate change negotiations. Thus, China was strongly opposed to it (MOC of China, 2009). A World Bank study by Mattoo *et al.* (2009) shows that such a carbon tariff would cut China's manufacturing exports by 21 percent and India's by 16 percent. No wonder that China and India warned angrily of trade wars if such border adjustment taxes were imposed (Reuters, 2009; *The Economist*, 2009).[25]

[23]President Obama was quoted as saying that "At a time when the economy worldwide is still deep in recession and we've seen a significant drop in global trade, I think we have to be very careful about sending any protectionist signals out there. I think there may be other ways of doing it than with a tariff approach." (Broder, 2009).

[24]The principle of "common but differentiated responsibilities" may be interpreted as permitting developing countries to move at different speeds of climate action relative to their developed counterparts. The aforementioned carbon tariff proposal required developing countries to take climate actions comparable to developed countries and thus runs counter to this principle.

[25]China's stance on carbon tariffs is in conflict with its statement for importers

6.3 The effectiveness and methodological challenges of border carbon adjustments

Proponents of an EAR argue that such a threat would be effective as an inducement for major emerging economies to take on such a level of climate action at which U.S. legislation aims. However, this is questionable. The EAR under the U.S. proposed cap-and-trade regime would not apply to all imports. Rather, it would specifically target primary emission-intensive products, such as steel, aluminium, and cement. Indeed, China has become a key producer of these primary products, accounting for 36% of global steel production, 32% of global aluminium production and over 50% of global cement production in 2007. The logic for the threat of EAR is that the fear of losing market access for these products would be enough to jawbone China to take climate actions that it otherwise would not take. However, the problem with this logic is that China's burgeoning supply of these carbon-intensive products is not mainly destined for export. Rather, they are made in China for China, going primarily to meet China's own demand. As the world's largest steel export, China exported only 2% of its steel production to the European Union and less than 1% to the U.S. in 2007. As the world's largest cement producer[26] and exporter, China consumed 97% of its cement domestically, and exported less than 1% of its production to the U.S. in 2007 (Houser, 2008; Houser *et al.*, 2008). Even if an EAR is implemented jointly with the European Union, it

being responsible for China's carbon emissions embodied in trade (Zhang, 2012). Being the workshop of the world and having the export-driven economy have led to a chunk of China's carbon emissions embedded in trade (e.g., Davis and Caldeira, 2010; Peters *et al.*, 2011). China certainly wants importers to cover some, if not all, of the costs of that. However, if this consumption-based accounting of CO_2 emissions, either implicitly or explicitly, is to indicate that the responsibility for the CO_2 emissions from the production of traded goods and services lies with the consumers in importing countries, it can then be argued that the final responsibility for regulating those CO_2 emissions lies with the governments of importing countries. Given that most carbon-intensive industries in the U.S. run a substantial trade deficit, this proposed EAR clearly aimed to level the carbon playing field for domestic producers and importers.

[26] In 2012, China produced 2.16 billion tons of cement, accounting for 58% of the total world cement production, and the India, the second largest producer, was far behind with 7% of the world production (ICR, 2013).

has little leverage effect on China because China is unlikely to raise the cost of producing 97% of its output for domestic market in order to protect a market of less than 3% of its production abroad.

Moreover, this effect on the targeted country will be further alleviated by re-routing trade flows to deliver the covered products from countries that are not subject to the EAR scheme. With Japan passing the comparability test and thus being exempted from an EAR under the proposed U.S. cap-and-trade regime, imposing an EAR on Chinese steel, but not on Japanese steel, could make Japanese steel more competitive in the U.S. market than Chinese steel. That could lead Japanese steel makers to sell more steel to the United States and Japanese steel consumers to import more from China (Houser *et al.*, 2008).[27] In the end, this neither affects China nor protects U.S. steel producers.

In the meantime, this undermines the effectiveness of the U.S. proposed carbon tariffs. Based on the multi-sector dynamic CGE model for China, Tang *et al.* (2015) show that, assuming a tax level of 60 US$/tC imposed jointly by both U.S. and EU, China's exports to U.S. and EU will be declined by about 0.79% and 1.33%, respectively, while the exports to Japan and rest of the world will be increased by about 0.38% and 0.48%, respectively, in 2020. Böhringer *et al.* (2010) show that China's and India's exports of the energy-intensive products are higher with the EU and the U.S. taking output-based allocations, exports rebates and import tariffs to complement their unilateral policy to cut their carbon emissions by 20% relative to their baselines than without any climate policies. Even if China's and India's exports of these energy-intensive products are lower when the EU and the U.S. take full border adjustment than without any climate policy, these sectors do not lose, because their aggregate production remains higher than without any climate policy.

[27] Kortum and Weisbach (2016) show that the re-routing problem is more complex than this simple evasion, taking a three-country case in which two counties (say, country H and country F. H does not impose BTAs on imports from F) price carbon and the third country (say, country T) does not have a carbon price. It can end up that no change has occurred in the patterns of production or consumption, but the border tax has been eliminated because the non-regulating country T exports its goods to the regulating country F and the regulating country H imports only goods from F instead of importing goods from both F and G initially.

Besides the issue of WTO consistency and the ineffectiveness of an EAR in leveraging developing countries to change behavior, there will be methodological challenges in implementing an EAR under a cap-and-trade regime, although such practical implementation issues are secondary concerns. Identifying the appropriate carbon contents embodied in traded products will present formidable technical difficulties, given the wide range of technologies in use around the world and very different energy resource endowments and consumption patterns among countries.

In the absence of any information regarding the carbon content of the products from exporting countries, importing countries, the U.S. in this case, could adopt either of the following two approaches to overcoming information challenges in practical implementation. One is to prescribe the tax rates for the imported product based on U.S. domestically predominant method of production for a like product, which sets the average embedded carbon content of a particular product (Zhang, 1998b; Zhang and Assunção, 2004). This practice is by no means without foundation. For example, the U.S. Secretary of the Treasury has adopted the approach in the tax on imported toxic chemicals under the Superfund Tax (GATT, 1987; Zhang, 1998b). An alternative is to set the best available technology (BAT) as the reference technology level and then use the average embedded carbon content of a particular product produced with the BAT in applying border carbon adjustments (Ismer and Neuhoff, 2007).

Generally speaking, developing countries will bear a lower cost based on either of the approaches than using the nation-wide average carbon content of imported products for the country of origin, given that less energy-efficient technologies in developing countries produce products of higher embedded carbon contents than those like products produced by more energy-efficient technologies in the U.S.. Mattoo *et al.* (2009) show that taxing the carbon footprint of imports based on U.S. domestic production would reduce China's and India's exports by around 3%, instead of 21% for China and 16% for India if a tariff is based on the amount emitted to make the imported products in China and India. However, to be more defensible, either of the approaches should allow foreign producers to challenge the carbon contents applied to their products to ensure that they will not pay for more than they have actually emitted. This is like carbon tariffs are targeted towards specific foreign producers, rather than based on some average embodied carbon content. Böhringer *et al.* (2017a) find that firm-targeted tariffs can deliver considerably stronger leakage reduction. They show that when a tariff design effectively targets the specific indirect emissions from the firm's use of electricity, the carbon leakage rate drops to 11.4% from the leakage rate of 20.2.% under the scenario without carbon tariffs, whereas the leakage rate drops to only 17.0% when tariffs are based on average regional embodied indirect emissions from electricity production. Moreover, firm-targeted tariffs benefit the exporters because the exporters are able to reduce their carbon tariffs by adjusting emissions, so that their competitiveness and the overall welfare of their economies will be less adversely affected than in the case of industry-level carbon tariff regimes. Overall, firm-targeted tariffs also deliver considerably higher gains in global cost-effectiveness than tariff designs at the industry level. Böhringer *et al.* (2017b) show that global welfare costs in the case of targeting specific indirect emissions from the firm's use of electricity can reduce by as much as 16% relative to the scenario without carbon tariffs.

7

Conclusions

In response to potentially severe climate change consequences, the OECD countries, in particular the EU, have taken the lead in cutting their greenhouse gas emissions. In the meantime, under the UNFCCC principle of "common but differentiated responsibilities," developing countries are allowed to move at different speeds relative to their developed counterparts. This difference in climate abatement commitments would persist at least in a Kyoto-type world until 2020. Under the Paris Agreement, almost all countries have outlined their climate action plans, but their climate mitigation targets outlined in terms of NDCs are very disparate in their ambition levels. Thus, while this post-Paris world looks different from the Kyoto-type world, the carbon leakage concern is reduced somewhat after the Paris Agreement but does not go away. Consequently, fragmented carbon markets and different carbon prices among trading partners will continue. Given the global nature of greenhouse gas emissions, the environmental effectiveness of the regulating country's more ambitious national climate efforts will thus be undermined through carbon leakage as domestic emission-intensive, trade-exposed industries would relocate to countries or regions without or with only quite lenient carbon pricing. As would be expected, the environmental effectiveness of such unilateral or uneven climate policies has long been academic focuses and of significant policy concerns even in the post-Paris world.

A large body of *ex ante* studies have quantified leakage impacts of unilateral or uneven climate policies at both the aggregated and sectoral levels. While one estimate reports an aggregated carbon leakage rate above 100% assuming the homogeneity of goods and increasing returns to scale, which differ from the usual assumptions of products differentiation and constant returns to scale under the CGE modeling, the estimated carbon leakage rates at the aggregated level tend to be much smaller than those estimates at sectoral levels. This should come as no surprise, because energy-intensive industries are major carbon emitters and are thus targeted by carbon abatement policies. Many studies have shown that these energy-intensive industries in the carbon constrained countries would suffer substantial economic losses and shift their investment and production to non- or less carbon constrained countries. Such a shifting will lead to high carbon leakages and thus reduce the environmental effectiveness of the carbon policy.

The fears of competitiveness losses and the environmental ineffectiveness as a result of carbon leakage are the main arguments put forward by industry lobby groups against stringent climate policies. Moreover, because potential costs are felt by these concentrated, often well-organized industries in comparison with widely-spread benefits, they are highly mobilized politically to exert greater influence on political negotiations and policy formulation. To mitigate their fears and gain their acceptance, the regulating countries often identify sectors deemed to be exposed to a risk of carbon leakage and grant these energy-intensive, trade exposed sectors some favorable treatments in the forms of exceptions from carbon pricing or free allowance allocations. This explains why *ex ante* estimates of potential carbon leakage rates are in sharp contrast with *ex post* results of environmental tax reforms and greenhouse gas emissions trading schemes that have been implemented in the EU. Indeed, the ETRs had not been significant in terms of their impact on unit production costs (below 1% in 50 cases out of 56 cases considered). While there is some evidence for a decline in competitiveness in selected countries/sectors, there is no consistent pattern and it is not possible to conclude that the reform was a significant contributing factor. Analysis of cement, iron and steel, aluminum and refinery sectors does not reveal carbon leakage for these trade-exposed carbon-intensive sectors during the first and second phase of the EU ETS. There is also no evidence that the EU ETS in place has triggered changes in trade flows and production patterns in these sectors during these phases.

While these *ex post* analyses of the EU ETS help us better understand competitiveness and carbon leakage impacts of the EU ETS, the insights from such analyses are of limited value for the future, because the EU ETS will become a much stricter scheme with a rising share of auctioning on the one hand and a decreasing yearly amount of the overall allowances handed out to industries on the other hand in the third phase 2013-2020 and beyond as the result of tightening the greenhouse gas reduction target and the ETS emissions cap itself. Moreover, the European Commission has implanted back-loading auction as a short term measure and has established the Market Stability Reserve as a structural measure, respectively, to address the current large accumulated allowance surplus. To accelerate the reduction of the current surplus of allowances in the EU carbon market, the intake rate (namely, a percentage rate of surplus allowances to be removed from the market and then to be placed in the MSR) of the previously agreed MSR will be doubled. As would be expected, the costs of emitting carbon are therefore expected to rise in the carbon constrained EU or other countries that take comparable climate actions, relative to the rest of the non- or less carbon constrained world.

To level the carbon costs in order to avoid putting industries exposed to a risk of carbon leakage at a competitive disadvantage vis-à-vis their trading partners, free output-based allocations and border carbon adjustments have been put forward. The effectiveness of both policies depends on the objective, namely, limiting carbon leakages or mitigating production loss in the sectors affected, and differs across sectors and countries. Studies differ when it comes to which anti-leakage policy, border adjustments or output-based allocation, is more effective to mitigating production loss, but tend to agree that border carbon adjustments are more effective than output-based allocation to tackle carbon leakage, although the two anti-leakage measures could achieve the equivalent effects when supplementing combining output-based rebating with a consumption tax for emission-intensive and tradeexposed goods.

However, Nordhaus (2015) argues that the core problem in designing a more robust international agreement is free riding to get the benefit of global emissions reductions while keeping lower energy costs for domestic use, and that leakage is just a minor aspect of free riding because most energy use is for domestic consumption. Therefore, Nordhaus has instead proposed to impose a uniform-percentage tariff on all imports from nonpricing countries. He argues that this would be far simpler than BCAs and yet better designed to address the free-riding problem. Another alternative to BCAs is to threaten to impose BCAs in the future on countries that do not meet a specified standard for emissions reductions or a specified carbon price. So, if producers know that imports from a given country will in the future face BCAs, they may not relocate there in the first place, reducing leakage (Kortum and Weisbach, 2016).

Even without consideration of these alternatives, however, the implementation of border carbon adjustment measures faces the effectiveness and methodological challenges. The sectors in developing countries specifically targeted by such anti-leakage measures do not necessarily lose, relative to a world without any climate polices. Moreover, the implementation of such measures would have potential conflicts with World Trade Organization provisions, depending on how they are designed and the specific conditions for implementing them (WTO and UNEP, 2009). Thus, in designing such trade measures, WTO rules need to be carefully scrutinised, and efforts need to be made early on to ensure that the proposed measures comply with them. After all, a conflict between the trade and climate regimes, if it breaks out, helps neither trade nor the global climate. Moreover, to increase the prospects for a successful WTO defence of the Waxman-Markey type of border adjustment provision, there should be: 1) a period of good faith efforts to reach agreements among the countries concerned before imposing such trade measures; 2) consideration of alternatives to trade provisions that could reasonably be expected to fulfill the same function but are not inconsistent or less inconsistent with the relevant WTO provisions; and 3) trade provisions that can refer to the designated special international reserve allowance pool, but should allow importers to submit equivalent emission reduction units that are recognized by international treaties to cover the carbon contents of imported products.

However, such anti-leakage policy, even if made WTO-consistent, is at most a short-term solution. Going forward, the global approach is the best option to level the carbon costs. The ideal solution is to have a legally binding international agreement that covers all countries, at least all major emitting economies, and sets their emissions limits. Such an agreement will enable to internalize the carbon costs and establish a global carbon price framework, thus leveling the carbon playing field. In this regard, the Paris Agreement is the significant step forward. This climate agreement specifies the long-term ambition of "holding the increase in the global average temperature to well below 2° above pre-industrial levels". However, the decentralized nature of the Paris Agreement and the requirements for countries to strengthen the levels of climate ambition over time in line with their national circumstances, combined with wide-ranging climate-related policies that countries with differentiated legal and political systems may pursue, would well lead to large asymmetries in shadow carbon pricing across countries, and undermine the environmental effectiveness of this international climate agreement. Clearly, establishing global sectoral agreements to supplement the Paris Agreement constitutes a robust policy to enhance carbon mitigation effect. Such cooperative sectoral agreements aim at either engaging specific sectors in non- or less carbon constrained countries and eliminating, at least reducing carbon-cost differences among countries of uneven climate policies, thus restoring a level carbon playing field. They can cover those sectors whose emissions are not covered by international climate agreements, such as emissions from fuels used for international aviation and maritime transport. To what extent global sectoral agreements can address competitiveness and leakage concerns depends on specific forms that such agreements take. This is the area where the carbon regulating countries need to explore, with their trading partners, to advance concerted climate mitigation efforts in a given sector at the international level. Unlike industry-led sectoral agreements, global sectoral agreements on both sectoral emissions caps and sector-specific performance standards and national sectoral binding emissions targets are subject to the UNFCCC negotiations. Provided that they are supported and enabled by finance and technology and are built upon a strong commitment by governments and the risk of a lowest common denominator can be avoided, these stringent forms of global sectoral agreements will supplement the Paris Agreement to address competitiveness and leakage concerns.

Acknowledgments

This monograph grows from the earlier version of my lengthy article "Competitiveness and Leakage Concerns and Border Carbon Adjustments" appearing in *International Review of Environmental and Resource Economics*, Vol. 6, No. 3, 2012, pp. 225-287. I would like to dedicate the monograph to my mother Guo Xié. She brought up all of her six children who all graduated from the Chinese universities, and passed away in 2012 when the agreement with the publisher NOW was signed for inclusion of the monograph in *Foundations and Trends in Microeconomics*. I like to acknowledge financial support by the National Natural Science Foundation of China (grant Nos. 71690243 and 71373055). An earlier draft of the monograph was presented at a plenary session on carbon pricing across the globe at the 35th International Association for Energy Economics International Conference, Perth, Australia, 24-27 June 2012. Part of the monograph formed a keynote address at the 3rd Congress of the East Asian Association of Environmental and Resource Economics, 20-22 February 2013. The monograph also formed part of the presentations at economics seminars at the National Development and Reform Commission, the Chinese Academy of Sciences, the Chinese Academy of Social Sciences, Peking University, Tsinghua University, Fudan University, Renmin University, Nankai University, and University of International Business and Economics. It has benefited from useful comments from three anonymous referees. That said, the views expressed here are those of the author. The author bears sole responsibility for any errors and omissions that may remain.

References

Abbasov, F. (2017). "Does the EU Have What it Takes to Be a Climate Leader?" *Huffington Post.* 31 May: Available at https://www.huffingtonpost.com/entry/does-the-eu-have-what-it-takes-to-be-a-climate-leader_us_592c310fe4b08861ed0ccb0c?utm_hp_ref=european-union.

Agnolucci, P. (2009). "The Effect of the German and British Environmental Taxation Reforms: A Simple Assessment". *Energy Policy.* 37: 3043–3051.

Aldy, J. E. and W. A. Pizer (2009). "The U.S. Competitiveness Impacts of Domestic Greenhouse Gas Mitigation Policies". Pew Center on Global Climate Change. Arlington.

Alexeeva-Talebi, V. (2011). "Cost Pass-through of the EU Emissions Allowances: Examining the European Petroleum Markets". *Energy Economics.* 33(S1): S75–S83.

Ambec, S., M. A. Cohen, S. Elgie, and P. Lanoie (2011). "The Porter Hypothesis at 20: Can Environmental Regulation Enhance Innovation and Competitiveness?". Discussion Paper 11-01, Resources for the Future, Washington, DC.

Andersen, M. S., T. Barker, E. Christie, P. Ekins, J. Fitz Gerald, J. Jilkova, J. Junankar, M. Landesmann, H. Pollitt, R. Salmons, S. Scott, and S. Speck (2007). "Competitiveness Effects of Environmental Tax Reforms (COMETR)". Final Report to the European Commission, DG Research and DG TAXUD, National Environmental Research Institute, University of Aarhus, Denmark, Available at http://www2.dmu.dk/cometr/COMETR_Final_Report.pdf.

Andersen, M. S. and P. Ekins, eds. (2009). *Carbon-energy Taxation: Lessons from Europe*. New York: Oxford University Press.

Anderson, B. and C. di Maria (2011). "Abatement and Allocation in the Pilot Phase of the EU ETS". *Environmental and Resource Economics*. 48(1): 83–103.

Armington, P. (1969). "A Theory of Demand for Products Distinguished by Place of Production". *IMF Staff Papers*. 16: 159–178.

Averchenkova, A. and S. Matikainen (2016). "Assessing the Consistency of National Mitigation Actions in the G20 with the Paris Agreement". Policy Brief, Grantham Research Institute on Climate Change and the Environment, London.

Babiker, M. H. (2005). "Climate Change Policy, Market Structure, and Carbon Leakage". *Journal of International Economics*. 65(2): 421–445.

Babiker, M. H. and T. F. Rutherford (2005). "The Economic Effects of Border Measures in Subglobal Climate Agreements". *Energy Journal*. 26(4): 99–126.

Barker, T., S. Junankar, and P. Summerton (2009). "The Effects of Environmental Tax Reform on International Competitiveness in the European Union: Modelling with E3ME". In: *Carbon-energy Taxation: Lessons from Europe*. M. S. Andersen and P. Ekins, eds., New York: Oxford University Press, pp. 147–214.

Baron, R. and ECON-Energy (1997). "Economic/Fiscal Instruments: Competitiveness Issues Related to Carbon/Energy Taxation". Policies and Measures for Common Action Working Paper14, Annex I Expert Group on the UNFCCC, OECD/IEA, Paris.

Baylis, K., D. Fullerton, and D. H. Karney (2014). "Negative Leakage". *Journal of the Association of Environmental and Resource Economists*. 1: 51–73.

Berger, J. R. (1999). "Unilateral Trade Measures to Conserve the World's Living Resources: An Environmental Breakthrough for the GATT in the WTO Sea Turtle Case". *Columbia Journal of Environmental Law*. 24: 355–411.

Bernard, A. L. and M. Vielle (2009). "Assessment of European Union Transition Scenarios with a Special Focus on the Issue of Carbon Leakage". *Energy Economics*. 31(S2): S274–S284.

Bhagwati, J. and P. C. Mavroidis (2007). "Is Action against US Exports for Failure to Sign Kyoto Protocol WTO-Legal?" *World Trade Review*. 6(2): 299–310.

Böhringer, C., B. Bye, T. Fæhn, and K. E. Rosendahl (2017a). "Targeted Carbon Tariffs: Export Response, Leakage and Welfare". *Resource and Energy Economics*. 50: 51–73.

Böhringer, C., C. Fischer, and K. E. Rosendahl (2010). "The Global Effects of Subglobal Climate Policies". *The B.E. Journal of Economic Analysis & Policy*. 10(2): Article 13.

Böhringer, C., C. Fischer, and K. E. Rosendahl (2014). "Cost-Effective Unilateral Climate Policy Design: Size Matters". *Journal of Environmental Economics and Management*. 67(3): 318–339.

Böhringer, C., K. E. Rosendahl, and H. B. Storrøsten (2017b). "Robust Policies to Mitigate Carbon Leakage". *Journal of Public Economics*. 149(2): 35–46.

Böhringer, C. and T. F. Rutherford (1997). "Carbon Taxes with Exemptions in an Open Economy: A General Equilibrium Analysis of the German Tax Initiative". *Journal of Environmental Economics and Management*. 32(2): 189–203.

Bollen, J., A. Gielen, and H. Timmer (1999). "Clubs, Ceilings and CDM: Macroeconomics of Compliance with the Kyoto Protocol". *Energy Journal*. 20: 177–206.

Bosetti, V., C. Carraro, E. Massetti, and M. Tavoni (2008). "International Energy R&D Spillovers and the Economics of Greenhouse Gas Atmospheric Stabilization". *Energy Economics*. 30(6): 2912–2929.

Bounds, A. (2006). "EU Trade Chief to Reject 'Green' Tax Plan". *Financial Times*, 17 December. Available at http://www.ft.com/cms/s/0/9dc90f34-8def-11db-ae0e-0000779e2340.html?nclick_check=1.

Bovenberg, A. L. and L. H. Goulder (2002). "Addressing Industry-Distributional Concerns in U.S. Climate Change Policy". Unpublished manuscript, Department of Economics, Stanford University.

Brack, D., M. Grubb, and C. Windram (1999). *International Trade and Climate Change Policies*. London: The Royal Institute of International Affairs and Earthscan.

Bradley, R., B. C. Staley, T. Herzog, J. Pershing, and K. A. Baumert (2007). "Slicing the Pie: Sector-Based Approaches to International Climate Agreements". World Resources Institute, Washington, DC.

Brandi, C. (2017). "Trade Elements in Countries' Climate Contributions under the Paris Agreement". International Centre for Trade and Sustainable Development, Geneva.

Branger, F. and P. Quirion (2013). "Understanding the Impacts and Limitations of the Current Instrument Mix in Detail: Industrial Sector". Technical Report, CIRED, France.

Branger, F. and P. Quirion (2014). "Would Border Carbon Adjustments Prevent Carbon Leakage and Heavy Industry Competitiveness Losses? Insights from a Meta-analysis of Recent Economic Studies". *Ecological Economics*. 99: 29–39.

Broder, J. (2009). "Obama Opposes Trade Sanctions in Climate Bill". *New York Times*. 28 June, Available at: http://www.nytimes.com/2009/06/29/us/politics/29climate.html?_r=2&scp=1&sq=obama%20opposes%20trade%20sanctions&st=cse.

Bruvoll, A. and T. Fæhn (2006). "Transboundary Effects of Environmental Policy: Markets and Emission Leakages". *Ecological Economics*. 59(4): 499–510.

Bruvoll, A. and B. M. Larsen (2004). "Greenhouse Gas Emissions in Norway: Do Carbon Taxes Work?" *Energy Policy*. 32(4): 493–505.

Buonanno, P., C. Carraro, and G. Marzio (2003). "Endogenous Induced Technical Change and the Costs of Kyoto". *Resource and Energy Economics*. 25(1): 11–34.

Burniaux, J. and J. O. Martins (2000). "Carbon Emission Leakages: A General Equilibrium View". *Tech. rep.* Economics Department Working Paper 242, OECD, Paris.

Burniaux, J.-M., J. Chateau, R. Dellink, R. Duval, and S. Jamet (2009). "The Economics of Climate Change Mitigation: How to Build the Necessary Global Action in a Cost-Effective Manner". OECD Economics Department, Working Paper 701, OECD, Paris.

Burniaux, J.-M., J. Chateau, R. Duval, and S. Jamet (2008). "The Economics of Climate Change Mitigation: Policies and Options for the Future". Economics Department Working Paper 658, OECD, Paris.

Capoor, K. and P. Abrosi (2008). "State and Trends of the Carbon Market 2008". The World Bank, Washington, DC.

Carbone J. C., C. H. and T. F. Rutherford (2009). "The Case for International Emission Trade in the Absence of Cooperative Climate Policy". *Journal of Environmental Economics and Management*. 58(3): 266–280.

Cement Sustainability Initiative (CSI) and European Cement Research Academy (ECRA) (2017). "Development of State of the Art-Techniques in Cement Manufacturing: Trying to Look Ahead". CSI/ECRA Technology Papers 2017, Dusseldorf and Geneva, March. Available at http://www.wbcsdcement.org/pdf/technology/CSI_ECRA_Technology_Papers_2017.pdf.

Charnovitz, S. (2003). "Trade and Climate: Potential Conflicts and Synergies". In: *Beyond Kyoto – Advancing the International Effort Against Climate Change*. Arlington: Pew Center on Global Climate Change, pp. 141–170.

Climate Strategies (2013). "Carbon Control Post 2020 in Energy Intensive Industries. Learning from the Experience in the Cement Sector". Energy-Intensive Industries Project Report, Climate Strategies, Cambridge, UK.

Clò, S. (2010). "Grandfathering, Auctioning and Carbon Leakage: Assessing the Inconsistencies of the New ETS Directive". *Energy Policy*. 38(5): 2420–2430.

Compston, H. and I. Bailey (2016). "Climate Policy Strength Compared: China, the US, the EU, India, Russia, and Japan". *Climate Policy*. 16(2): 145–164.

Convery, F., D. Ellerman, and C. de Perthuis (2008). "The European Carbon Market in Action: Lessons from the First Trading Period". Interim Report, Report No. 162, The MIT Joint Program on the Science and Policy of Global Change, Cambridge, MA.

Copeland, B. R. and M. S. Taylor (2003). *Trade and the Environment: Theory and Evidence*. Princeton: Princeton University Press.

Copeland, B. R. and M. S. Taylor (2005). "Free Trade and Global Warming: A Trade Theory View of the Kyoto Protocol". *Journal of Environmental Economics and Management*. 49(2): 205–234.

Copeland, B. R. and S. Taylor (1994). "North-South Trade and the Environment". *Quarterly Journal of Economics*. 109: 755–87.

Davis, S. J. and K. Caldeira (2010). "Consumption-based Accounting of CO_2 Emissions". *Proceedings of National Academy of Sciences*. 107(12): 5687–5692.

de Cendra, J. (2006). "Can Emissions Trading Schemes Be Coupled with Border Tax Adjustments? An Analysis vis-à-vis WTO Law". *Review of European Community and International Environmental Law (RECIEL)*. 15(2): 131–145.

Demailly, D. and P. Quirion (2006). "CO_2 Abatement, Competitiveness and Leakage in the European Cement Industry under the EU ETS: Grandfathering versus Output-based Allocation". *Climate Policy*. 6(1): 93–113.

Demailly, D. and P. Quirion (2008). "Leakage from Climate Policies and Border Tax Adjustment: Lessons from a Geographic Model of the Cement Industry". In: *The Design of Climate Policy*. R. Guesnerie and H. Tulkens, eds., Cambridge: The MIT Press, pp. 333–358.

Dissou, Y. and T. Eyland (2011). "Carbon Control Policies, Competitiveness, and Border Tax Adjustments". *Energy Economics*. 33: 556–564.

Dong, Y. and J. Whalley (2009a). "Carbon Motivated Regional Trade Arrangements: Analytics and Simulations". NBER Working Paper No. 14880, National Bureau of Economic Research, Cambridge, MA.

Dong, Y. and J. Whalley (2009b). "How Large Are the Impacts of Carbon Motivated Border Tax Adjustments". NBER Working Paper No. 15613, National Bureau of Economic Research, Cambridge, MA.

Dröge, S., H. van Asselt, T. Brewer, M. Grubb, R. Ismer, Y. Kameyama, M. Mehling, S. Monjon, K. Neuhoff, P. Quirion, K. Schumacher, L. Mohr, W. Suwala, Y. Takamura, T. Voituriez, and X. Wang (2009). "Tackling Leakage in a World of Unequal Carbon Prices". Synthesis Report. Climate Strategies, Cambridge, UK.

Earth Negotiations Bulletin (ENB) (2017). "COP23 Summary of the Fiji / Bonn Climate Change Conference". 6-17 November. Available at http://enb.iisd.org/download/pdf/enb12714e.pdf.

Ederington, J., A. Levinson, and J. Minier (2005). "Footloose and Pollution-Free". *Review of Economics and Statistics*. 87(1): 92–99.

Ederington, J. and J. Minier (2003). "Is Environmental Policy a Secondary Trade Barrier? An Empirical Analysis". *Canadian Journal of Economics*. 36(1): 137–154.

Ekins, E. and R. Salmons (2007). "An Assessment of the Impacts of ETR on the Competitiveness of Selected Industrial Sectors". In: *Competitiveness Effects of Environmental Tax Reforms* (COMETR), M. S. Andersen et al., COMETR WP3, pp. 24–35.

Ekins, P. (2012). "Sustainable Growth Revisited: Technology, Economics and Policy". *Mineral Economics*. 24(2/3): 59–71.

Ellerman, A. D. and B. K. Buchner (2007). "The European Union Emissions Trading Scheme: Origins, Allocations, and Early Results". *Review of Environmental Economics and Policy*. 1(1): 66–87.

Ellerman, A. D. and B. K. Buchner (2008). "Over-Allocation or Abatement? A Preliminary Analysis of the EU ETS Based on the 2005-06 Emissions Data". *Environmental and Resource Economics*. 41(2): 267–287.

Ellerman, A. D., F. J. Convery, C. de Perthuis, E. Alberola, B. K. Buchner, A. Delbosc, C. Hight, J. H. Keppler, and F. C. Matthes (2010). *Pricing Carbon: The European Union Emissions Trading Scheme*. Cambridge: Cambridge University Press.

Ellerman, A. D., P. L. Joskow, J. P. Montero, R. Schmalensee, and E. M. Bailey (2000). *Markets for Clean Air: The US Acid Rain Program*. Cambridge: Cambridge University Press.

Elliott, J., I. Foster, S. Kortum, T. Munson, F. P. Cervantes, and D. Weisbach (2010). "Trade and Carbon Taxes". *American Economic Review*. 100(2): 465–469.

Enevoldsen, M. K., A. V. Ryelund, and M. S. Andersen (2007). "Decoupling of Industrial Energy Consumption and CO_2-emissions in Energy-intensive Industries in Scandinavia". *Energy Economics*. 29(4): 665–692.

European Commission (2008). "Proposal for a Directive of the European Parliament and of the Council Amending Directive 2003/87/EC so as to Improve and Extend the Greenhouse Gas Emission Allowance Trading System of the Community". COM(2008) 16 final, Brussels.

European Commission (2009a). "Directive 2009/29/EC of the European Parliament and of the Council of 23 April 2009 Amending Directive 2003/87/EC so as to Improve and Extend the Greenhouse Gas Emission Allowance Trading Scheme of the Community (Text with EEA Relevance)". Available at http://eur-lex.europa.eu/LexUriServ/LexUriServ.do?uri=OJ:L:2009:140:0063:0087:EN:PDF.

European Commission (2009b). "Impact Assessment Accompanying Document to the Commission Decision Determining a List of Sectors and Subsectors Which Are Deemed to Be Exposed to a Significant Risk of Carbon Leakage Pursuant to Article 10a (13) of Directive 2003/87/EC". C(2009)10251 final, Draft Commission Staff Working Document, Brussels, 24 December. Available at, http://ec.europa.eu/clima/policies/ets/leakage/docs/sec_2009_1710_en.pdf.

European Commission (2011a). "Emissions Trading: Questions and Answers on Rules and Guidance on Allocation of Free Allowances to the Power Sector". MEMO/11/201, Brussels, 29 March. Available at http://europa.eu/rapid/pressReleasesAction.do?reference=MEMO/11/201.

European Commission (2011b). "Commission Decision of 27 April 2011 Determining Transitional Union-wide Rules for Harmonised Free Allocation of Emission Allowances Pursuant to Article 10a of Directive 2003/87/EC of the European Parliament and of the Council, Notified under Document C(2011) 2772, 2011/278/EU. Available at http://eur-lex.europa.eu/LexUriServ/LexUriServ.do?uri=OJ:L:2011:130:0001:0045:EN:PDF.

European Commission (2012a). Analysis of Options beyond 20% GHG Emission Reductions: Member State Results,Staff Working Paper, Brussels, February.

European Commission (2012b). Q&A Emissions Trading: Commission Prepares to Change the Time Profile for Auctions of Emission Allowances, MEMO/12/600, 25 July. Available at http://europa.eu/rapid/pressReleasesAction.do?reference=MEMO/12/600.

European Commission (2012c). "Information Provided on the Functioning of the EU Emission Trading System, the Volumes of Greenhouse Gas Emission Allowances Auctioned and Freely Allocated and the Impact on the Surplus of Allowances in the Period up to 2020". Commission Staff Working Document (provisional version), Brussels, July.

European Parliament and the Council (2008). "Directive of the European Parliament and of The Council Amending Directive 2003/87/EC to Enhance Cost-effective Emission Reductions and Low-Carbon Investments and Decision (EU) 2015/1814". Brussels, 14 February. Available at http://data.consilium.europa.eu/doc/document/PE-63-2017-INIT/en/pdf.

European Parliament and the Council (2015). "Decision (EU) 2015/1814 of the European Parliament and of the Council of 6 October 2015 Concerning the Establishment and Operation of a Market Stability Reserve for the Union Greenhouse Gas Emission Trading Scheme and Amending Directive 2003/87/EC". Brussels, October Available at https://publications.europa.eu/en/publication-detail/-/publication/01c4f171-6e49-11e5-9317-01aa75ed71a1/language-en.

Felder, S. and T. Rutherford (1993). "Unilateral CO_2 Reductions and Carbon Leakage: The Consequences of International Trade in Oil and Basic Materials". *Journal of Environmental Economics and Management*. 25(2): 162–176.

Fischer, C. and A. Fox (2009). "Comparing Policies to Combat Emissions Leakage: Border Tax Adjustments versus Rebates". Discussion Paper 09–02, Resources for the Future, Washington, DC.

Garnaut, R. (2008). *The Garnaut Climate Change Review*. Cambridge: Cambridge University Press.

Genasci, M. (2008). "Border Tax Adjustments and Emissions Trading: The Implications of International Trade Law for Policy Design". *Carbon and Climate Law Review*. 2(1): 33–42.

General Agreement on Tariffs and and Trade (GATT) (1987). "United States – Taxes on Petroleum and Certain Imported Substances". Report of the Panel, Adopted on 17 June, L/6175, BISD 34S/136, Geneva Available at http://www.wto.org/english/res_e/booksp_e/analytic_index_e/introduction_01_e.htm.

General Agreement on Tariffs and Trade (GATT) (1990). "Thailand – Restrictions on Importation of and Internal Taxes on Cigarettes". Report of the Panel, DS10/R, Adopted on 7 November, BISD 37S/200, Geneva Available at http://www.wto.org/english/res_e/booksp_e/analytic_index_e/introduction_01_e.htm.

General Agreement on Tariffs and Trade (GATT) (1994). "United States: Restrictions on the Imports of Tuna". Report of the Panel (not adopted), DS29/R, 16 June, Geneva. Available at http://www.wto.org/english/res_e/booksp_e/analytic_index_e/introduction_01_e.htm.

Gerlagh, R. and O. Kuik (2007). "Carbon Leakage with International Technology Spillovers". Working Papers 2007.33, Fondazione Eni Enrico Mattei, Italy.

Gielen, D. J. and Y. Moriguchi (2002). "CO_2 in the Iron and Steel Industry: An Analysis of Japanese Emission Reduction Potentials". *Energy Policy*. 30(10): 849–863.

Gillingham, K., R. G. Newell, and W. A. Pizer (2008). "Modeling Endogenous Technological Change for Climate Policy Analysis". *Energy Economics*. 30(6): 2734–2753.

Golombek, R. and M. Hoel (2004). "Unilateral Emission Reductions and Cross-Country Technology Spillovers". *Advances in Economic Analysis & Policy*. 4(2): Article 3.

Goulder, L. H. and R. N. Stavins (2011). "Interactions between State and Federal Climate Change Policies". NBER Working Paper 16123. National Bureau of Economic Research, Cambridge, MA.

Grossman, G. (1980). "Border Tax Adjustment: Do They Distort Trade?" *Journal of International Economics*. 10(1): 117–128.

Grossman, G. M. and A. B. Krueger (1993). "Environmental Impacts of a North American Free Trade Agreement". In: *The Mexico-U.S. Free Trade Agreement*. eds., P.M. Garber (ed.), Cambridge: MIT Press, pp. 13–56.

Grubb, M. (2012). "Strengthening the EU ETS - Creating a Stable Platform for EU Energy Sector Investment". Climate Strategies, Cambridge, UK.

Grubb, M., C. Hope, and R. Fouquet (2002). "Climatic Implications of the Kyoto Protocol: The Contribution of International Spillover". *Climatic Change*. 54(1/2): 11–28.

Guangdong Provincial Development and Reform Commission (GPDRC) (2013). "A Circular on the Initial Carbon Allowance Allocation and Work Plan of Guangdong". November 25 Available at: http://www.gddpc.gov.cn/xxgk/tztg/201311/t20131126_230325.htm.

Haverkamp, J. (2008). "International Aspects of a Climate Change Cap and Trade Program". Testimony before the Committee on Finance, U.S. Senate, 14 February. Available at https://www.finance.senate.gov/imo/media/doc/021408jhtest1.pdf.

Hoel, M. (1996). "Should a Carbon Tax Be Differentiated across Sectors?" *Journal of Public Economics*. 59(1): 17–32.

Höhne, N., H. Fekete, M. G. J. den Elzen, A. F. Hof, and T. Kuramochi (2017). "Assessing the Ambition of Post-2020 Climate Targets: A Comprehensive Framework". *Climate Policy*.

Hollinger, P. (2009). "Sarkozy Renews Carbon Tax Call". *Financial Times*, 11 September: p. 5.

Hourcade, J.-C., D. Demailly, K. Neuhoff, and M. Sato (2007). "Differentiation and Dynamics of EU ETS Competitiveness Impacts". Climate Strategies, Cambridge, UK.

Houser, T. (2008). "Carbon Tariffs – Why Trade Sanctions Won't Work". *China Economic Quarterly*. 12(3): 33–38.

Houser, T., R. Bradley, B. Childs, J. Werksman, and R. Heilmayr (2008). "Leveling the Carbon Playing Field: International Competition and U.S. Climate Policy Design". Peterson Institute for International Economics and World Resources Institute, Washington, DC.

Howse, R. (2010). "Climate Mitigation Subsidies and the WTO Legal Framework: A Policy Analysis". International Institute for Sustainable Development, Winnipeg, Canada.

Hubei Provincial Government (HPG) (2014). "Interim Administrative Measures for Carbon Emissions Trading in Hubei Province". April 4 Available at: http://gkml.hubei.gov.cn/auto5472/auto5473/201404/t20140422_497476.html.

Hufbauer, G. and J. Kim (2009). "Climate Policy Options and the World Trade Organization". *Economics-eJournal.* Article 3: 2009–29.

ICR (2013). "The Global Cement Report 10th Edition". International Cement Review, London.

Intergovernmental Panel on Climate Change (IPCC) (2001). *Climate Change 2001: Mitigation.* Working Group III Contribution to the Third Assessment Report. Cambridge: Cambridge University Press.

Intergovernmental Panel on Climate Change (IPCC) (2007). *Climate Change 2007: Mitigation of Climate Change.* Working Group III Contribution to the Fourth Assessment Report, Cambridge: Cambridge University Press.

Intergovernmental Panel on Climate Change (IPCC) (2014). *Climate Change 2014: Mitigation of Climate Change.* Working Group III Contribution to the Fourth Assessment Report, Cambridge: Cambridge University Press.

International Carbon Action Partnership (ICAP) (2018). *Emissions Trading Worldwide: Status Report 2018.* Berlin.

International Civil Aviation Organization (ICAO) (2016a). "Resolution A39-2: Consolidated Statement of Continuing ICAO Policies and Practices Related to Environmental Protection – Climate Change". Montreal October. Available at https://www.icao.int/environmental-protection/Documents/Resolution_A39_2.pdf.

International Civil Aviation Organization (ICAO) (2016b). "Resolution A39-3: Consolidated Statement of Continuing ICAO Policies and Practices related to Environmental Protection – Global Market-based Measure (MBM) Scheme". Montreal, October. Available at https://www.icao.int/environmental-protection/Documents/Resolution_A39_3.pdf.

Ismer, R. and K. Neuhoff (2007). "Border Tax Adjustment: A Feasible Way to Support Stringent Emission Trading". *European Journal of Law and Economics.* 24(2): 137–164.

Jaffe, A., R. Newell, and R. Stavins (2002). "Environmental Policy and Technological Change". *Environmental and Resource Economics.* 22: 41–69.

Jaffe, A., P. Peterson, P. Portney, and R. Stavins (1995). "Environmental Regulation and the Competitiveness of US Manufacturing: What Does the Evidence Tell Us?" *Journal of Economic Literature.* 33(1): 132–163.

James, S. (2009). "A Harsh Climate for Trade: How Climate Change Proposals Threaten Global Commerce". Trade Policy Analysis No. 41, CATO Institute Center for Trade Policy Studies, Washington, DC.

Junginger, M. (2005). *Learning in Renewable Energy Technology Development.* Ph.D Dissertation, The Netherlands: Utrecht University.

Junginger, M., W. van Sark, and A. Faaij, eds. (2010). *Technological Learning in the Energy Sector.* Cheltenham, UK and Northampton, USA: Edward Elgar.

Kallbekken, S. (2007). "Why the CDM will Reduce Carbon Leakage". *Climate Policy.* 7(3): 197–211.

Kallbekken, S., L. Flottorp, and N. Rive (2007). "CDM Baseline Approaches and Carbon Leakage". *Energy Policy.* 35(8): 4154–4163.

Kazmerski, L. L. (2011). "Solar Photovoltaics: No Longer an Outlier". In: *Global Dynamics in the Green Energy Industry: A New Engine of Growth.* F. Fesharaki, N. Y. Kim, Y. H. Kim and Z. X. Zhang, eds., Seoul: Korean Energy Economics Institute Press, pp. 48–80.

Kettner, C., D. Kletzan-Slamanig, and A. Köppl (2012). "The EU Emission Trading Scheme: National Allocation Patterns and Trading Flows". Working Paper, Austrian Institute of Economic Research, Vienna.

Kettner, C., D. Kletzan-Slamanig, and A. Köppl (2013). "The EU Emission Trading Scheme – Sectoral Allocation Patterns and Factors Determining Emission Changes". Working Paper No. 444, Austrian Institute of Economic Research, Vienna.

Kortum, S. and D. A. Weisbach (2016). "Border Adjustments for Carbon Emissions: Basic Concepts and Design". *Tech. rep.* Discussion Paper No. 16-09, Resources for the Future, Washington, DC.

Krugman, P. (2009b). "Carbon Tariffs". *The New York Times*, 19 July. Available at http://krugman.blogs.nytimes.com/2009/07/19/carbontariffs/?scp=3&sq=carbon%20tariffs%20krugman&st=cse..

Krugman, P. (1994). "Competitiveness: A Dangerous Obsession". *Foreign Affairs*. 73(2): 28–44.

Krugman, P. (2009). "Climate, Trade, Obama". *The New York Times*, 29 June. Available at: http://krugman.blogs.nytimes.com/2009/06/29/climate-trade-obama/.

Lessmann, K., R. Marschinski, and O. Edenhofer (2009). "The Effects of Tariffs on Coalition Formation in a Dynamic Global Warming Game". *Economic Modelling*. 26: 641–649.

Levinson, A. and M. S. Taylor (2008). "Unmasking the Pollution Haven Effect". *International Economic Review*. 49(1): 223–254.

Löschel, A. (2002). "Technological Change in Economic Models of Environmental Policy: A Survey". *Ecological Economics*. 43(2/3): 105–126.

Majocchi, A. and M. Missaglia (2002). "Environmental Taxes and Border Tax Adjustment". Societ Italiana Economisti Pubblici (SIEP) Working Paper No. 127.

Markusen, J. R. (1975). "International Externalities and Optimal Tax Structures". *Journal of International Economics*. 5: 15–29.

Martin, R., M. Muûls, L. B. de Preux, and U. J. Wagner (2014). "Industry Compensation Under Relocation Risk: A Firm-Level Analysis of the EU Emissions Trading Scheme". *American Economic Review*. 104(8): 2482–2508.

Mathiesen, L. and O. Maestad (2004). "Climate Policy and the Steel Industry: Achieving Global Emission Reductions by an Incomplete Climate Agreement". *Energy Journal*. 25(4): 91–114.

Mattoo, A., A. Subramanian, D. van der Mensbrugghe, and J. He. (2009). *Reconciling Climate Change and Trade Policy*. Washington, DC, November: World Bank Working Paper No. 5123.

McBroom, M. (2008). "How the IBEW-UWM-Boilermakers-AEP International Proposal Operates within Climate Legislation." 17 June, Available at http://www.wita.org/index.php?tg=fileman&idx=viewfile&idf=189&id=4&gr=Y&path=&file=WITA-+Climate+Change+-+Overview+of+IBEWAEP+Proposal+(June+17%2C+2008).pdf.

McDonald, A. and L. Schrattenholzer (2001). "Learning Rates for Energy Technologies". *Energy Policy*. 29: 255–261.

Mckibbin, W. J. and P. J. Wilcoxen (2009). "The Economic and Environmental Effects of Border Tax Adjustments for Climate Policy". In: *Climate Change, Trade and Competitiveness: Is a Collision Inevitable? Brookings Trade Forum: 2008/2009*. L. Brainard and I. Sorkin, eds., Washington, DC: Brookings Institution Press.

McMackin, J. (2009). "Testimony for the Energy Intensive Manufacturers' Working Group on Greenhouse Gas Regulation before the House Committee on Ways and Means". 24 March.

Meckling, J. O. and G. Y. Chung (2009). "Sectoral Approaches to International Climate Policy: A Typology and Political Analysis". Discussion Paper 2009-02. Belfer Center for Science and International Affairs, JFK School of Government, Harvard University, Cambridge, MA.

Miltner, A. and R. Salmons (2009). "Trends in the Competitiveness of Selected Industrial Sectors in ETR Countries". In: *Carbon-energy Taxation: Lessons from Europe*. Andersen, M. S. and P. Ekins, eds., New York: Oxford University Press.

Ministry of Commerce of China (MOC of China) (2009). "A Statement on Carbon Tariffs". 3 July, Beijing, Available at http://www.mofcom.gov.cn/aarticle/ae/ag/200907/20090706375686.html (in Chinese).

Monjon, S. and P. Quirion (2011). "Addressing Leakage in the EU ETS: Border Adjustment or Output-based Allocation?" *Ecological Economics*. 70(11): 1957–1971.

Morris, D. (2012). "Losing the Lead? Europe's Flagging Carbon Market". Sandbag, London.

Morris, M. G. and E. D. Hill (2007). "Trade is the Key to Climate Change". *The Energy Daily*. 35(33).

Murphy, A. (2017). "MEPs Express Doubts over Global Aviation CO_2 Scheme While Reforming ETS". Transport & Environment, Brussels. 26 July, Available at: https://www.transportenvironment.org/news/meps-express-doubts-over-global-aviation-co2-scheme-while-reforming-ets.

Neuhoff, K. (2008). "Tacking Carbon: How to Price Carbon for Climate Policy". University of Cambridge, Available at http://www.eprg.group.cam.ac.uk/wp-content/uploads/2009/03/tackling-carbon_final_3009082.pdf.

Newell, R., A. Jaffe, and R. Stavins (1999). "The Induced Innovation Hypothesis and Energy-Saving Technological Change". *Quarterly Journal of Economics*. 114(3): 941–975.

Nordhaus, W. (2002). "Modeling Induced Innovation in Climate Change Policy". In: *Technological Change and the Environment*. A. Grubler, N. Nakicenovic, and W. Nordhaus, eds., Washington, DC: Resources for the Future Press.

Nordhaus, W. (2015). "Climate Clubs: Overcoming Free-Riding in International Climate Policy". *American Economic Review*. 105(4): 1339–1370.

Obergassel, W., C. Arens, L. Hermwille, N. Kreibich, F. Mersmann, H. Ott, and H. Wang-Helmreich (2016). "Phoenix from the Ashes: An analysis of the Paris Agreement to the United Nations Framework Convention on Climate Change". Wuppertal Institute for Climate, Environment and Energy, Wuppertal, Germany.

OECD (2003). "Environmental Policy in the Steel Industry: Using Economic Instruments". COM/ENV/EPOC/DAFFE/FCA(2002)68/FINAL, Paris. Available at: http://www.oecd.org/dataoecd/58/20/33709359.pdf.

OECD (2010). "Innovation Effects of the Swedish NOx Charge". OECD Environment Directorate, Paris.

Oikonomou, V., M. Patel, and E. Worrell (2006). "Climate Policy: Bucket or Drainer?" *Energy Policy*. 34(18): 3656–3668.

Parker, L. and J. Grimmett (2009). *Climate Change: EU and Proposed U.S. Approaches to Carbon Leakage and WTO Implications*. Report R40914, Congress Research Service, Washington, DC. Available at: https://digital.library.unt.edu/ark:/67531/metadc627109/m1/1/high_res_d/R40914_2009Nov04.pdf.

Parry, Ian W. H., R. C. Williams III, and L. H. Goulder (1999). "When Can Carbon Abatement Policies Increase Welfare? The Fundamental Role of Distorted Factor Markets". *Journal of Environmental Economics and Management*. 37(1): 52–84.

Pearce, D. (1991). "The Role of Carbon Taxes in Adjusting to Global Warming". *The Economic Journal*. 101: 938–948.

Peters, G. P., J. C. Minx, C. L. Weber, and O. Edenhofer (2011). "Growth in Emission Transfers via International Trade from 1990 to 2008". *Proceedings of the National Academy of Sciences*. 108(2): 8903–8908.

Pezzey, J. (1992). "Analysis of Unilateral CO_2 Control in the European Community and OECD". *The Energy Journal*. 13(3): 159–172.

Pizer, W. A. and D. Popp (2008). "Endogenizing Technological Change: Matching Empirical Evidence to Modeling Needs". *Energy Economics*. 30(6): 2754–2770.

Popp, D. (2002). "Induced Innovation and Energy Prices". *American Economic Review*. 92(1): 160–180.

Porter, M. (1991). "America's Green Strategy". *Scientific American*. 264(4): 168.

Porter, M. and C. van der Linde (1995). "Toward a New Conception of the Environment-Competitiveness Relationship". *Journal of Economic Perspective*. 9(4): 97–118.

Quirion, P. (2009). "Historic versus Output-based Allocation of GHG Tradable Allowances: A Comparison". *Climate Policy*. 9(6): 575–592.

Quirion, P. (2010). "Competitiveness and Leakage". In: *Climate Change Policies: Global Challenges and Future Prospects*. E. Cerdá and X. Labandeira, eds., Cheltenham, UK, and Northampton, MA: Edward Elgar, pp. 105–122.

Reinaud, J. (2008a). "Climate Policy and Carbon Leakage – Impacts of the European Emissions Trading Scheme on Aluminium". IEA Information Paper, IEA/OECD, Paris.

Reinaud, J. (2008b). "Issues behind Competitiveness and Carbon Leakage – Focus on Heavy Industry". IEA Information Paper, IEA/OECD, Paris. Available at https://www.iea.org/publications/freepublications/publication/Competitiveness_and_Carbon_Leakage.pdf.

Reuters (2009). "China Says "Carbon Tariffs" Proposals Breach WTO Rules". *New York Times*. 3 July.

Rivers, N. (2010). "Impacts of Climate Policy on the Competitiveness of Canadian Industry: How Big and How to Mitigate?" *Energy Economics*. 32(5): 1092–1104.

Rose, A., T. Peterson, and Z. X. Zhang (2006). "Regional Carbon Dioxide Permit Trading in the United States: Coalition Choices for Pennsylvania". *Penn State Environmental Law Review*. 14(2): 203–229.

Rose, A. and Z. X. Zhang (2004). "Interregional Burden-Sharing of Greenhouse Gas Mitigation in the United States". *Mitigation and Adaptation Strategies for Global Change*. 9(4): 477–500.

Rosendahl, K. E. and J. Strand (2011). "Carbon Leakage from the Clean Development Mechanism". *The Energy Journal*. 32(4): 27–50.

Rutherford, T. (1992). "The Welfare Effects of Fossil Carbon Restrictions: Results from a Recursively Dynamic Trade Model". OECD Economics Department Working Papers 112, OECD, Paris.

Saito, M. (2004). "Armington Elasticities in Intermediate Inputs Trade: A Problem in Using Multilateral Trade Data". *Canadian Journal of Economics*. 37(4): 1097–1117.

Sato, M., K. Neuhoff, V. Graichen, K. Schumacher, and F. Matthes (2015). "Sectors under Scrutiny: Evaluation of Indicators to Assess the Risk of Carbon Leakage in the UK and Germany". *Environmental and Resource Economics*. 60: 99–124.

Sijm, J. P. M., O. J. Kuik, M. Patel, V. Oikonomou, E. Worrell, P. Lako, E. Annevelink, G. J. Nabuurs, and H. W. Elbersen (2004). "Spillovers of Climate Policy: An Assessment of the Incidence of Carbon Leakage and Induced Technological Change due to CO_2 Abatement Measures". *Tech. rep.* Report 500036002, The Netherlands Research Programme on Climate Change, Bilthoven, The Netherlands.

Sijm, J., K. Neuhoff, and Y. Chen (2006). "CO_2 Cost Pass through and Windfall Profits in the Power Sector". *Climate Policy.* 6(1): 49–72.

Smale, R., M. Hartley, C. Hepburn, J. Ward, and M. Grubb (2006). "The Impact of CO_2 Emissions Trading on Firm Profits and Market Prices". *Climate Policy.* 6(1): 31–48.

Smith, S. (2008). "Environmentally Related Taxes and Tradable Permit Systems in Practice". Environment Directorate, OECD, Paris.

Söderholm, P. and G. Klaassen (2007). "Wind Power in Europe: A Simultaneous Innovation–diffusion Model". *Environmental and Resource Economics.* 36: 163–190.

Söderholm, P. and T. Sundqvist (2007). "Empirical Challenges in the Use of Learning Curves for Assessing the Economic Prospects of Renewable Energy Technologies". *Renewable Energy.* 32: 2559–2578.

Speck, S. (2007). "Carbon Leakage". In: *Competitiveness Effects of Environmental Tax Reforms.* (COMETR). Andersen, M. S. et al., COMETR WP5, pp. 47–56.

Stiglitz, J. E. (2006). "A New Agenda for Global Warming". *The Economists' Voice.* 3(7): Article 3.

Subcommittee on Energy and Air Quality of the U.S. House of Representatives (2008). "Competitiveness Concerns/Engaging Developing Countries". Climate Change Legislation Design White Paper, Washington DC, January. Available at http://energycommerce.house.gov/Climate_Change/White_Paper.competitiveness.013108.pdf.

Summerton, P., H. Pollit, and G. Klaassen (2015). "A Model-Based Assessment of First-Mover Advantage and Climate Policy". *Environmental Economics and Policy Studies.* 17(2): 299–312.

Swedish National Board of Trade (2004). *Climate and Trade Rule – Harmony or Conflict?* Stockholm.

Tang, L., Q. Bao, Z. X. Zhang, and S. Wang (2015). "Carbon-based Border Tax Adjustments and China's International Trade: Analysis based on a Dynamic Computable General Equilibrium Model". *Environmental Economics and Policy Studies.* 17(2): 329–360.

The Economist (2008a). "Pollution Law: Trading Dirt". 7 June, pp. 42–44.

The Economist (2008b). "Emissions Suspicions". 19 June, Available at: http://www.economist.com/finance/displaystory.cfm?story_id=11581408.

The Economist (2009). "Green with Envy: The Tension between Free Trade and Capping Emissions". 21 November, p. 80.

The World Bank (2007). *International Trade and Climate Change: Economic, Legal and Institutional Perspectives.* Washington, DC.

U. S. House of Representatives (2009). "H.R. 2998". 23 June. Available at: http://thomas.loc.gov/cgi-bin/query/z?c111:H.R.2998.

United Nations Conference on Trade and Development (UNCTAD) (2010). *Trade and Environment Review 2009/2010: Promoting Poles of Clean Growth to Foster the Transition to a More Sustainable Economy.* UNCTAD/DITC/TED/2009/2, New York and Geneva.

United Nations Framework Convention on Climate Change (UNFCCC) (2015). "Adoption of the Paris Agreement". FCCC/CP/2015/L.9/Rev.1, Twenty-First Session of the Conference of the Parties, Paris, 30 November – 11 December.

U.S. Environmental Protection Agency (EPA) (2009). "Comparison of FTI, EPA Analyses of H.R. 2454, Title IV". Memorandum to the House Energy, and Commerce Committee Staff, Washington, DC, 10 June.

Veenendaal, P. and T. Manders (2008). "Border Tax Adjustment and the EU-ETS: A Quantitative Assessment". Central Planning Bureau (CPB) Document No.171, The Hague.

Wigle, R. (2001). "Sectoral Impacts of Kyoto Compliance". *Tech. rep.* Report No. 34, Industry Canada, Ottawa, ON.

Winchester, N., S. Paltsev, and J. M. Reilly (2011). "Will Border Carbon Adjustments Work?" *The B.E. Journal of Economic Analysis & Policy.* 11(1): Article 7.

World Trade Organization (WTO) (1998). "United States – Import Prohibition of Certain Shrimp and Shrimp Products". Report of the Appellate Body, WT/DS58/AB/R, Geneva.

World Trade Organization (WTO) (2001). "United States – Import Prohibition of Certain Shrimp and Shrimp Products, Recourse to Article 21.5 of the DSU by Malaysia". Panel Report, WT/DS58/RW, Adopted on 21 November, Geneva.

World Trade Organization (WTO) and United Nations Environment Programme (UNEP) (2009). *Trade and Climate Change: WTO-UNEP Report*. Geneva.

WSA (2012). "World Steel in Figures 2012". Technical Report, World Steel Association, Brussels.

Wu, B. (2014). "The State Council establishes the Beijing-Tianjin-Hebei Collaborative Development Leading Group". *Southern Metropolis Daily*. 3 August: Available at http://news.ifeng.com/a/20140803/41419143_0.shtml?src=se6_newtab.

Zhang, Z. K. and Z. X. Zhang (2017). "Intermediate Input Linkage and Carbon Leakage". *Environment and Development Economics*. 22(6): 725–746.

Zhang, Z. X. (1998a). *The Economics of Energy Policy in China: Implications for Global Climate Change*. New Horizons in Environmental Economics Series, Cheltenham, UK and Northampton, USA: Edward Elgar.

Zhang, Z. X. (1998b). "Greenhouse Gas Emissions Trading and the World Trading System". *Journal of World Trade*. 32(5): 219–239.

Zhang, Z. X. (1999). "Should the Rules of Allocating Emissions Permits be Harmonised?" *Ecological Economics*. 31(1): 11–18.

Zhang, Z. X. (2000). "Can China Afford to Commit itself an Emissions Cap? An Economic and Political Analysis". *Energy Economics*. 22(6): 587–614.

Zhang, Z. X. (2004). "Open Trade with the U.S. without Compromising Canada's Ability to Comply with its Kyoto Target". *Journal of World Trade*. 38(1): 155–182.

Zhang, Z. X. (2006). "Cutting Carbon Emissions While Making Money: A Wishful Thinking or a Win-win Opportunity?" The Keynote Address at the Plenary Session on Making Money from Saving Carbon at the 29th International Association for Energy Economics International Conference. 8-10 June, Potsdam, Germany.

Zhang, Z. X. (2007). "Why Has China not Embraced a Global Cap-and-Trade Regime?" *Climate Policy*. 7(2): 166–170.

Zhang, Z. X. (2008). "Asian Energy and Environmental Policy: Promoting Growth While Preserving the Environment". *Energy Policy*. 36: 3905–3924.

Zhang, Z. X. (2009). "Multilateral Trade Measures in a Post-2012 Climate Change Regime?: What Can Be Taken from the Montreal Protocol and the WTO?" *Energy Policy*. 37: 5105–5112.

Zhang, Z. X. (2010). "Is It Fair to Treat China a Christmas Tree to Hang Everybody's Complaints? Putting Its Own Energy-saving into Perspective". *Energy Economics*. 32: S47–S56.

Zhang, Z. X. (2012). "Who Should Bear the Cost of China's Carbon Emissions Embodied in Goods for Exports?" *Mineral Economics*. 24(2/3): 103–117.

Zhang, Z. X. (2014). "Energy Prices, Subsidies and Resource Tax Reform in China". *Asia and the Pacific Policy Studies*. 1(3): 439–454.

Zhang, Z. X. (2015a). "Crossing the River by Feeling the Stones: The Case of Carbon Trading in China". *Environmental Economics and Policy Studies*. 17(2): 263–297.

Zhang, Z. X. (2015b). "Carbon Emissions Trading in China: The Evolution from Pilots to a Nationwide Scheme". *Climate Policy*. 15: S104–S126.

Zhang, Z. X. (2017). "Are China's Climate Commitments in a Post-Paris Agreement Sufficiently Ambitious?" *Wiley Interdisciplinary Reviews: Climate Change*. 8(2): e443.

Zhang, Z. X. and L. Assunção (2004). "Domestic Climate Policy and the WTO". *The World Economy*. 27(3): 359–386.

Zhang, Z. X. and A. Baranzini (2004). "What Do We Know About Carbon Taxes? An Inquiry into Their Impacts on Competitiveness and Distribution of Income". *Energy Policy*. 32(4): 507–518.

Zhang, Z. X. and A. Nentjes (1998). "International Tradeable Carbon Permits as a Strong Form of Joint Implementation". In: *Pollution for Sale: Emissions Trading and Joint Implementation*. eds., S. Sorrel and J. Skea, Cheltenham, England: Edward Elgar, pp. 322–342.

Lightning Source UK Ltd.
Milton Keynes UK
UKHW021424090221
378492UK00005B/1263

9 781680 834826